Lecture Notes in Networks and Systems

Volume 297

The series "Lecture Notes in Networks and Systems" publishes the latest developments in Networks and Systems—quickly, informally and with high quality. Original research reported in proceedings and post-proceedings represents the core of LNNS.

Volumes published in LNNS embrace all aspects and subfields of, as well as new challenges in, Networks and Systems.

The series contains proceedings and edited volumes in systems and networks, spanning the areas of Cyber-Physical Systems, Autonomous Systems, Sensor Networks, Control Systems, Energy Systems, Automotive Systems, Biological Systems, Vehicular Networking and Connected Vehicles, Aerospace Systems, Automation, Manufacturing, Smart Grids, Nonlinear Systems, Power Systems, Robotics, Social Systems, Economic Systems and other. Of particular value to both the contributors and the readership are the short publication timeframe and the world-wide distribution and exposure which enable both a wide and rapid dissemination of research output.

The series covers the theory, applications, and perspectives on the state of the art and future developments relevant to systems and networks, decision making, control, complex processes and related areas, as embedded in the fields of interdisciplinary and applied sciences, engineering, computer science, physics, economics, social, and life sciences, as well as the paradigms and methodologies behind them.

Indexed by SCOPUS, INSPEC, WTI Frankfurt eG, zbMATH, SCImago.

All books published in the series are submitted for consideration in Web of Science.

More information about this series at http://www.springer.com/series/15179

Karen Lizbeth Flores Rodríguez ·
Raymundo Ramos Alvarado · Masoud Barati ·
Veronica Segovia Tagle ·
Roberto Sostrand Velázquez González
Editors

Recent Trends in Sustainable Engineering

Proceedings of the 2nd International Conference on Applied Science and Advanced Technology

Springer

Editors
Karen Lizbeth Flores Rodríguez
CICATA-IPN
Instituto Politécnico Nacional
Querétaro, Mexico

Masoud Barati
Cardiff University
Cardif, UK

Roberto Sostrand Velázquez González
CICATA-IPN
Instituto Politécnico Nacional
Querétaro, Mexico

Raymundo Ramos Alvarado
CICATA-IPN
Instituto Politécnico Nacional
Querétaro, Mexico

Veronica Segovia Tagle
CICATA-IPN
Instituto Politécnico Nacional
Querétaro, Mexico

ISSN 2367-3370 ISSN 2367-3389 (electronic)
Lecture Notes in Networks and Systems
ISBN 978-3-030-82063-3 ISBN 978-3-030-82064-0 (eBook)
https://doi.org/10.1007/978-3-030-82064-0

This Springer imprint is published by the registered company Springer Nature Switzerland AG
The registered company address is: Gewerbestrasse 11, 6330 Cham, Switzerland

Organization

Organizing Committee

General Chair

Karen Flores, Instituto Politécnico Nacional, MX

Program Chair

Angel Hernández, Instituto Politécnico Nacional, MX

Technical Program Chair

Raymundo Ramos, Instituto Politécnico Nacional, MX

Conference Treasure Chair

Alan Lugo, Instituto Politécnico Nacional, MX

Local Arrangements Chair

Bruno Flores, Universidad Autónoma de Querétaro, MX

Publicity Chair

Dagoberto Pulido, Instituto Politécnico Nacional, MX

Technical Program Committee

Masoud Barati, Edinburgh Napier University, UK
Araceli López, Instituto Politécnico Nacional, MX
Verónica Segovia, Instituto Politécnico Nacional, MX
Roberto Velázquez, Instituto Politécnico Nacional, MX

Publicity Committee

Marisol Alvarado, Instituto Politécnico Nacional, MX
Pedro Martínez, Instituto Politécnico Nacional, MX
Lucero Flores, Instituto Politécnico Nacional, MX

Cultural Committee

Kemisseth Marcos, Instituto Politécnico Nacional, MX
Mayra Cándido, Instituto Politécnico Nacional, MX

Collaborators

Araceli Zapatero, Universidad Anahuac, MX
Eddy Olmedo, Instituto Politécnico Nacional, MX
Marycarmen Feregrino, Instituto Politécnico Nacional, MX
Abraham López, Instituto Politécnico Nacional, MX
Gloria Girón, Instituto Politécnico Nacional, MX
Daniel Canales, Instituto Politécnico Nacional, MX
Guillermo Luque, Instituto Politécnico Nacional, MX
Yuriria Acuña, Xiafra, MX

Advisory Board

Daniela Calvo, Universidad Nacional Autónoma de México, MX
Juan Hurtado, Instituto Politécnico Nacional, MX
Edith Muñoz, Instituto Politécnico Nacional, MX

Scientific Committee

Eduardo Morales Sánchez, Instituto Politécnico Nacional, MX
Xochitl Yamile Sandoval Castro, Instituto Politécnico Nacional, MX
Juvenal Rodriguez Resendiz, Universidad Autónoma de Querétaro, MX
Iván Trejo Zuñiga, Universidad Tecnológica de San Juan del Rio, MX
Alonso Alejandro Jiménez Garibay, Tecnológico Nacional de México en Celaya, MX
Mario Enrique Duarte Gonzalez, Universidad Antonio Nariño, CO
Antonio Hernández Zavala, Instituto Politécnico Nacional, MX
José Aníbal Arias, Universidad Tecnológica de la Mixteca, MX
Juan José Martínez Nolasco, Tecnológico Nacional de México en Celaya, MX
Araceli Zapatero Gutierrez, Universidad Anahuac, MX
Carlos Velasco Santos, Tecnológico Nacional de México, MX
Carolina Hernández Navarro, Tecnológico Nacional de México, MX
Rafael Martínez Martínez, Universidad Tecnológica de la Mixteca, MX
Linda Viviana García Quiñonez, CICESE Unidad Monterrey, MX
Eduardo Arturo Elizalde Peña, Universidad Autónoma de Querétaro, MX
Aníbal Uriel Pacheco Sánchez, Universitat Autònoma de Barcelona, ES
Iván Domínguez López, Instituto Politécnico Nacional, MX
Frank Otremba, Federal Institute for Materials Research and Testing, Germany
José Manuel Robles Solís, Universidad Politécnica de Zacatecas, MX

Daniela Kristell Calvo Ramos, Universidad Nacional Autónoma de México, MX
Jesús Antonio Camarillo Montero, Universidad Veracruzana, MX
Rafael Peña Gallardo, Universidad Autónoma de San Luis Potosí, MX
José Porfirio González Farías, Instituto Nacional de Mexico en Celaya, MX
Nildia Yamileth Mejias Brizuela, Universidad Politécnica de Sinaloa, MX
María Cristina Castañon Bautista, Universidad Autónoma de Baja California, MX
Daniel Humberto Solís Recéndez, UPIIZ-IPN, MX
Alicia Ravelo, Universidad Autónoma de Baja California, MX
María Andrade Aréchiga, Universidad de Colima, MX
Francisco J. Hernandez-Lopez, CIMAT-Mérida, MX
Víctor Manuel Zamudio Rodríguez, Tecnológico Nacional de México en León, MX
Geovanni Martínez, Universidad de Costa Rica
Masoud Barati, Edinburgh Napier University, UK
Fernando Ireta Muñoz, I3S/CNRS, France
José Joel Gonzalez Barbosa, Instituto Politécnico Nacional, MX
Felipe de Jesús Trujillo Romero, Universidad de Guanajuato, MX
Alfonso Padilla Vivanco, Universidad Politécnica de Tulancingo, MX
Carina Toxqui Quitl, Universidad Politécnica de Tulancingo, MX
Dagoberto Pulido Arias, Instituto Politécnico Nacional, MX
Angel Moises Hernandez Ponce, Instituto Politécnico Nacional, MX
Othón González Chavéz, Instituto Politécnico Nacional, MX
María Valentina Angoa Peréz, CIIDIR-IPN-MICHOACAN, MX
Karol Karla García Aguirre, UPIIZ-IPN, MX
Carolina Estefanía Chávez Murillo, UPIIZ-IPN, MX
Elsa Verónica Herrera Mayorga, Universidad Autónoma de Tamaulipas, MX
Manuel Vazquez Vazquez, Universidad de Santiago de Compostela, ES
Marcela Gaytán Martínez, Universidad Autónoma de Querétaro, MX
Pedro Alberto Vazquez Landaverde, Instituto Politécnico Nacional, MX
Claudia Gutiérrez Antonio, Universidad Autónoma de Querétaro, MX
Ericka Santacruz Juárez, Universidad Politécnica de Tlaxcala, MX
Genaro Iván Cerón Montes, Universidad Tecnológica de Tecámac, MX
Iván Luzardo Ocampo, Universidad Autónoma de Querétaro, MX
Juan L. Silva, Mississippi State University, USA
M. Patricia Santiago Gómez, Universidad Tecnológica de la Mixteca, MX
Eric Ortega Sánchez, Universidad Popular Autónoma del Estado de Puebla, MX

Sponsors

Instituto Politécnico Nacional
Centro de Investigación en Ciencia Aplicada y Tecnología Avanzada
SMM Capítulo estudiantil CFATA UNAM Juriquilla
Consejo de Ciencia y Tecnología del Estado de Querétaro
Spectro Networks

Secretaría de Turismo Querétaro
EcoMaker Store
SerInge
Chocolates Truffel
3D THEA & Nebo
Joyería Científica Son Para Químicos
The Geek Squirrel
No Soy Nerd
Diseñoños
La ruta del becario
Bittbuk

Preface

The book in here is a compilation of works presented during the 2nd International Conference on Applied Science and Advanced Technology (iCASAT 2021), which was organized by students, researchers, and administrative of the Centro de Investigación en Ciencia Aplicada y Tecnología Avanzada Unidad Querétaro in collaboration with the student chapter of the Sociedad Mexicana de Materiales (SMM) from the Centro de Física Aplicada y Tecnología Avanzada of UNAM Juriquilla.

For this occasion, we met in a virtual modality due to the current sanitary conditions. However, instead of being a disadvantage, it was an opportunity to connect and share our knowledge with people that in another situation, we could not be together.

The book is a multidisciplinary space and serves as a platform to share and learn about the frontier knowledge between different areas related to "Recent trends in sustainable engineering."

Sustainable engineering promotes the responsible use of resources and materials involved in the different manufacturing processes or the execution stages of a service. An interdisciplinary approach is required in all aspects of engineering. In this sense, engineers, researchers, and the academic community will play a fundamental role in developing new technologies that respect the environment, still, at the same time, that considers social and economic factors.

We thank the authors of the articles for their trust in choosing us as a platform to share their work. We can imagine that they may not have the results wanted on the first try. However, instead of giving up, they keep working in their minds, laboratories, and workshops to be the best they can. Also, we are grateful to the scientific committee members for taking the time to contribute their experience to advise and correct the published works.

Organizing Committee

Contents

Development of Virtual Router Machine for Modbus Open Connection

Heber Hernández-Vázquez, Irma Y. Sanchez, Fernando Martell, Jose E. Guzman, and Raul A. Ortiz

Abstract Virtual machines enable high performance of corresponding physical machines, and also analysis, design and testing of the associated process and control systems. Interaction with virtual machines allows training as well as efficient operation of a running productive process; the former is the scope of this contribution for automation design and implementation abilities for remote education, especially important in times of encouraged social distance due to the current pandemic situation. This work presents the development of a virtual router from its 3D graphics to its preparation for online training purposes. The Modbus communication protocol is used to make the signals of the virtual machine available to a costumed control system, giving the characteristic of open connection. The virtual router instruments a virtual remote laboratory for designing, and testing automation systems proposed by interested users. The procedure for the creation of this virtual machine is detailed using industrial and open-access tools such as Blender® and Modbus TCP. The resulting virtual machine is presented and its potential use in a virtual laboratory is proposed. Further work is directed to the development of a cyberphysical system for the router.

H. Hernández-Vázquez · I. Y. Sanchez (✉) · J. E. Guzman · R. A. Ortiz
Universidad Politécnica de Aguascalientes, Aguascalientes, México
e-mail: isanchez@ingmt.com; irma.sanchez@upa.edu.mx

H. Hernández-Vázquez
e-mail: mc190007@alumnos.upa.edu.mx

J. E. Guzman
e-mail: jose.guzman@upa.edu.mx

R. A. Ortiz
e-mail: raul.ortiz@upa.edu.mx

I. Y. Sanchez
Ingeniería Mecatrónica, S.A., Aguascalientes, México

F. Martell
Centro de Investigaciones en Óptica, Aguascalientes, México
e-mail: fmartell@cio.mx

K. L. Flores Rodríguez et al. (eds.), *Recent Trends in Sustainable Engineering*, Lecture Notes in Networks and Systems 297,
https://doi.org/10.1007/978-3-030-82064-0_1

Keywords Virtual twin · Modbus TCP · Virtual laboratory · Online training

1 Introduction

In industrial automation, the optimization of the production processes relies heavily on the implementation of advanced control functions. Support personnel of the manufacturing systems must develop skills in diverse conventional automation technologies, such as programmable controllers and industrial communication protocols, in order to be able to adopt new automation technologies. Industry 4.0 is characterized by a higher level of interaction with processes, enabling access to information and operation of plant equipment from remote locations, under the goal of integrated manufacturing [1]. Cyberphysical systems constitute an important technological trend to achieve these characteristics. The virtual twin of a physical machine is an essential element of a cyberphysical production system to enhance process and operation efficiency and safety, even in set-up or commissioning procedures [2, 3].

The convenience of the semantic description of cyberphysical systems has been stated given the rapid technology changes in instrumentation and controllers. According to the metamodel of a cyberphysical system exposed in [4], the contribution of this work is the connection of the virtual entity, as an addition of the physical machine or process to be controlled and monitored, to the computational system where the user algorithms are defined in the chosen control platform. The Modbus protocol, being a current industrial communication trend, gives the intended general open connectivity for the developed virtual machine.

Researchers have worked on the evaluation of Modbus communication based on a software master device to detect data transmission errors before system deployment or delivery [5]. In the context of cyberphysical systems, the Modbus protocol has been used for the implementation of a supervisory control and data acquisition system (SCADA) for a planar three-degree-of-freedom robotic manipulator [6]. The system incorporated communication fault detection by the Hamming codification technique. The position achieved by the virtual manipulator was passed to the real manipulator, that is, specific elements were communicated for a predefined task. The scope of the present work with a virtual router is an open system to be closed by the developer's application and instrumentation.

The configuration of a PLC of a particular brand (RSLogix 5000) to communicate to a virtual robot in a virtual laboratory has been addressed [7], using Modbus TCP to ease the transition to real industrial equipment, both as for transferring trained competences as well as for exporting control solutions. Here, the article focuses on the preparation of the virtual machine to be communicated with practically any PLC by considering a soft-PLC.

A digital or virtual twin is defined as a virtual representation of the dynamics of a physical system with data transfer that allows proper operation and monitoring of the latter [8]. The proper representation of the physical system for a cloud-based cyberphysical system encompasses multiple layers of data transfer and information generation, such as collection of data from the physical twin and its environment, processing and transmitting data to different production areas, mathematical algorithms and CAD models, use of data for prediction, optimization or control, and visualization through a graphical user interface [9]. However, a preliminary version, in the sense of development of only essential elements of data exchange and use in different levels can also be considered as a digital twin, including the case of virtual model with proper dynamics but no information flow with the real space [8].

Applications of virtual twins range from products and processes to company assets and services [8]. The role of virtual twins can be as a resource for monitoring the integrity of the physical system, tracking the life cycle of a product (design, service, recycling or disposal), supporting decision making during design and operation, simulating the real counterpart, and fast and effective optimization of processes based on synchronization and prediction of the seamless connected physical system [10].

The characteristics of a virtual twin are the following: having a corresponding physical entity, definition of physical and virtual environments (the other real process that interact with the physical twin, and other computer systems that use the virtual twin), fidelity to the physical system comprising quality, quantity, and values of parameters, unidirectional and bidirectional connection between physical and virtual twins, and synchronization between them, and with other processes [11].

Different authors have outlined frameworks for the development of a digital twin based on product traits and functions, production system requirements, and service value creation. For a determined physical product, the general steps for creating a digital twin are (1) building a tridimensional representation using CAD tools, (2) data collection and analysis including implementation of learning techniques, (3) simulation of product behaviors, (4) adjusting real product behavior based on recommendations from simulations, (5) stablishing real-time bidirectional communications between real and virtual twin product, (6) gather product related information from different sources, mainly customers by different means [12]. In the case of a production line, the digital twin development comprises the definition of (1) plant layout, (2) functions for each process unit, and (3) real parameters for process cycles [13]. A complex digital twin for service value creation must reflect the hierarchical structure of technology and business related to a product and leads to the concept of family of twins [14]. Conversely, the virtual twin of a product, production line and service can be used for the validation of designs of respective physical entities.

A virtual machine, separated from the physical entity for operation, gives many of the benefits of a virtual twin, those regarding off-line operation for control, product design, and training. The functionality of the virtual machine is a milestone in the development of a digital twin or, furthermore, of a cyberphysical system. Therefore, aside from a work in progress, the virtual machine constitutes a product itself that can be used as a remote and virtual laboratory for learning purposes [15, 16], particularly of automation engineers that may test control strategies for later training operators of such strategies. Additionally, situations of sanitary restrictions make a virtual machine an important tool for continuous and distance education [17]. In this work, although the terms of virtual twin and virtual machine can be interchanged, the virtual machine is defined as a system, whose inputs and outputs communicate with an arbitrary control platform, that is, open for the connection of an electronic control unit (ECU).

This paper is organized as follows: Sect. 2 describes the development steps of the virtual router emphasizing the characteristics of the real machine and used tools. Section 3 shows the 3D representation of the router, and explains how to configure its animation. Section 4 is dedicated to the dynamic modeling of the router actuators. Section 5 defines the information transfer using the Modbus protocol, and describes the performance of the virtual machine. Section 6 concludes this article by commenting the potential use of the virtual router, and future work.

2 Method and Resources

The objective of the proposed virtual router machine is creating an easily accessible working tool that provides a simulation environment for acquiring control design and programming abilities, or testing automation solutions before putting them into practice with the real router. Therefore, the virtual twin method steps are limited to the essential characteristics attributed to the virtual machine as a development phase of the virtual twin. The development of the virtual router is also based on commonly available and copyright free working tools.

2.1 Method for the Development of the Virtual Machine

The development method used for the virtual router machine is adapted from a virtual product twin based approach [8]. The steps of the procedure used to generate the virtual machine are the following:

1. Creation of the 3D graphic model of the router virtual twin.
2. Assembly and animation of the parts of the virtual twin that integrates a dynamic model of the router actuators.
3. Preparation of a bidirectional connection with the virtual twin by means of a communication protocol.

The concept of the use of the virtual machine to be developed is shown in Fig. 1. Graphics and open connection for the virtual machine derived from the proposed method steps are illustrated.

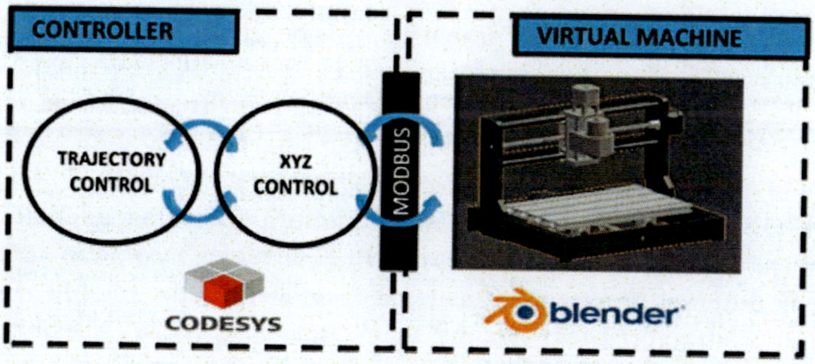

Fig. 1 Conceptual diagram of the interaction of the virtual machine with a controller

2.2 *Resources: Materials and Software Tools*

The physical machine is a commercial three-axis router model CNC3 3018 PRO, useful to work on wood, plastics, PVC, soft aluminum (6061), brass, PCB's, among other malleable materials. The machine dimensions are 330 × 400 × 240 mm (13.0′ × 15.7″ × 9.4″) and has an effective working area of 300 × 180 × 45 mm (11.8″ × 7.1″ × 1.8″). The main components of the CNC3 3018 PRO are shown in Table 1.

The software used to produce the tridimensional visual model includes SolidWorks® and Blender® v.2.79. SolidWorks® is a CAD design software for modeling 3D and 2D parts and assemblies, and Blender® is an open source cross-platform software for 3D modeling, animation, lighting, rendering and graphics creation. The 3D model of each of the pieces of the router was made in SolidWorks®, and then imported into Blender®, where the animation of the virtual twin was created with its game logic engine (Blender Game Engine, BGE). The complete graphical modeling of the router could have been made in Blender®, however familiarity with SolidWorks® was useful, as well as its compatibility with the Blender® platform.

Table 1 CNC3 3018 PRO part list

Part	Characteristics	Quantity
Aluminum profiles	20 × 40 × 290 mm	2
	20 × 20 × 360 mm	2
Aluminum workbench	180 × 300 mm	1
Guide rail (X axis)	10 × 360 mm	2
Guide rail (Y axis)	10 × 290 mm	2
Lead screw (X axis)	T8 * 4, 365 mm	1
Lead screw (Y axis)	T8 * 4, 297 mm	1
Stepper motor	NEMA17, 42 × 34 mm	3
Spindle motor and chuck	775-ER11	1
Control board	GRBL 1.1	1

Another feature of Blender® is that it allows programming using the Python language, which was used to program the animation.

The communication protocol selected for the virtual machine was Modbus TCP. This open source communication protocol is based on a client/server architecture and is supported by most industrial programmable controllers.

CODESYS® was used as a general platform to interact with the virtual machine. This platform has all the features of a PLC, in which the control logic can be created using different programming languages defined by the EIC61131 standard (ladder logic or structured text, among them), and also allows visualizations of system variables.

3 Graphical 3D Model

The 3D model of the virtual machine should resemble the real router. However, exactitude in the representation of small components is not required to mimic the movements of the machine, since the operation of the machine for testing control programs is the focus of this development.

3.1 Elements and Dimensions

The machine parts were drawn in SolidWorks® according to the general shape characteristics and selected exact dimensions of the router components. The parts were grouped according to the elements to be animated in the virtual machine in the X, Y and Z axes. The X and Z axes were assigned, respectively, to the horizontal and vertical directions for the movement of the router tool in the vertical plane; the worktable is moved along the Y axis. Figure 2 shows the assembly schemes in the X, Y and Z axes.

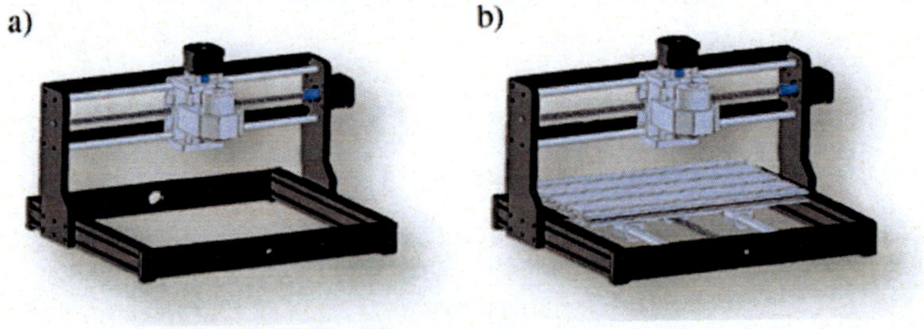

Fig. 2 Tridimensional model of the router: **a** assembly for the X and Z movement axes, **b** assembly of worktable to move along the Y axis

3.2 Animation

The assembly and animation of the virtual twin were done in Blender® software. The animation of the virtual router was made with the BGE. In the BGE logic module, the execution, debugging and execution of objects are performed within a video game platform using a wide range of dedicated functions, used in this case to control the X, Y and Z axes of the virtual machine. The configuration steps are explained next.

Step 1. Configure the logic editor panel, where the logic, properties and states to control the behavior of the virtual twin objects are set. The logic editor panel has 3 elements that must be configured to obtain the control logic for the animation: sensor, controller and actuator. This configuration is shown below:

1. The sensor is configured as an always type element: it gives a continuous output signal at regular intervals. The always mode is used for actions that must be done at every tick or at the start of program execution (Fig. 3).
2. The animation controller is configured as a Python script to be executed when a sensor trigger signal is received (Fig. 4).
3. The actuator motion type is set as simple motion. This configuration allows to control the position and velocity, but it does it as an instantaneous displacement (Fig. 5).

Step 2. Elaborate the Python script. The Python script can interact with the scene through the Blender® API (application programming interface). A Python script can

Fig. 3 Always sensor

Fig. 4 Python controller

Fig. 5 Motion actuator

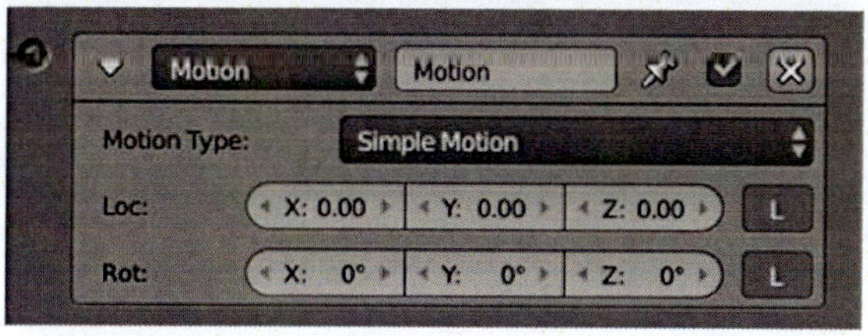

be run as a complete file or by single modules. To create a Python script, there is a special section in the BGE called Text Editor dedicated to write the code. The programming process requires importing pre-installed libraries from Blender® or installing complementary libraries by third parties. The algorithm to implement consists of the following sequence of steps:

1. Import module of motion logical functions from the BGE.
2. Get the addresses of the motion controller and actuators, and activate the motion module.
3. Produce motion of the animated object through X, Y and Z actuators.
4. Get the position and orientation of the animated object.

This algorithm is the basis for the animation of the virtual twin. The execution of the corresponding script performs the animation process in Blender® in 1 ms.

4 Dynamic Model

The movement dynamics along one axis is represented by a system with a pulse frequency percentage input and a position output. The shaft velocity is an internal variable determined by the applied input and integrated to give the total displacement. A time dynamics is adjusted to reproduce the desired transitory behavior. This modeling approach is illustrated by the block diagram of Fig. 6.

Although an instantaneous response of a stepper motor can be assumed, this modeling scheme allows the consideration of a transitory behavior, and can also be adapted to direct current motors.

The dynamics of the stepper motor motion is given by a first order recursive equation with a sample time of 100 ms, unit gain, and unit sample time to time constant ratio, and is shown in Eq. 1, where k is the discrete time:

$$m(k) = e^{-1}m(k-1) + \left(1 - e^{-1}\right)u(k-1) \tag{1}$$

The displacement produced is calculated by numerical integration according to Eq. 2, using the sample time (T), the operation parameter value for maximum pulse frequency (F_{max}), and motor parameters: number of steps (n_s), gear ratio (revolutions

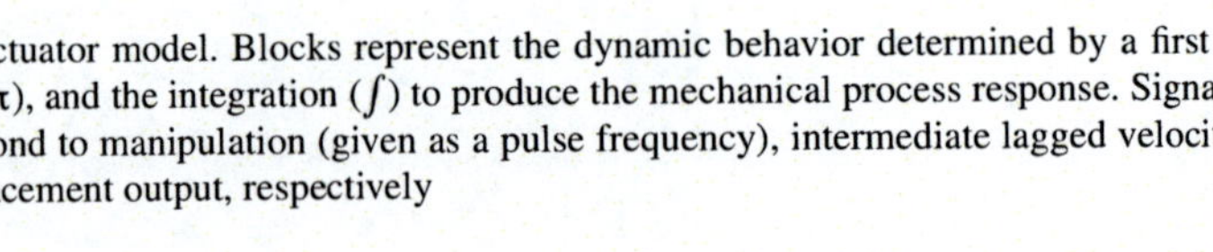

Fig. 6 Actuator model. Blocks represent the dynamic behavior determined by a first order time constant (τ), and the integration ($\int$) to produce the mechanical process response. Signals u, m and d correspond to manipulation (given as a pulse frequency), intermediate lagged velocity variable, and displacement output, respectively

of lead screw per revolution of motor shaft, r_g) and pitch (linear displacement per lead screw turn, p):

$$d(k) = d(k-1) + Tm(k)F_{max}\frac{1}{n_s}r_g p \tag{2}$$

For the router CNC3 3018 PRO, n_s=16, r_g=1, and $p = 2$ mm/rev.

The value of $d(k)$ indicates the position acquired along one axis, therefore emulates a sensor output. These equations are defined for each actuator on X, Y and Z axes within the Python script for the animation control program. The manipulation value $u(k)$ is to be determined by the control program elaborated by the user. Therefore, only the dynamics of the actuators and animation logic are programed on the side of the virtual machine.

5 Communication for Open Connection

Modbus is an industrial communication protocol that has been a “de facto” standard since its introduction as an open protocol. Today, Modbus is widely used by many automation technology manufacturers and is found in programmable controllers, operation panels and human machine interfaces, smart instruments, industrial robots and many other industrial devices and systems. Dominant automation brands such as Siemens and Allen Bradley have incorporated Modbus protocols among their connectivity options. Modbus is an application layer protocol that allows communications over RS-485, RS-232 or USB with the Modbus RTU version. The Modbus TCP version is a specification over Ethernet network TCP/IP protocols. The latter is used to prepare the connection of the virtual machine to a PLC.

5.1 Configuration of Server and Client

The bidirectional communication between Blender® and the CODESYS® platforms is established through the Modbus communication protocol. Blender® requires a library containing the Modbus open source code developed for Python called pymodbus. This library has the different Modbus communication modes (including the client/server architecture), as well as the different types of registers. The virtual machine in Blender® was configured as the client and CODESYS® as the server. The configuration and use of Blender® as a client in Python with the pymodbus library consists of the following steps:

1. Import Modbus client from pymodbus library.
2. Configure the logging format and get the client logger.
3. Define the Modbus client by address, port, and client number.

4. Connect the client for logging.
5. Read input registers using initial address, number of registers, and client number.
6. Write holding registers using address, data stream, and client number.
7. Log off the client.

The configuration of the virtual machine as client is more convenient because the CODESYS® SoftPLC needs to be configured as server, which is the role of PLCs when they interact with human–machine interfaces or remote terminal units in SCADA systems. This configuration leaves all the Modbus read and write messages to the virtual machine while the PLC or any other ECU can be dedicated mainly to control tasks. The Modbus server function is embedded in the PLC firmware, and often only a memory mapping is required to handle input/output data. This is a critical advantageous characteristic for the proposed open connection of the virtual machine.

The transferred variables are the manipulation variables for the three actuators, and the position values along the three axes.

5.2 *Modbus Data Communication Performance*

The following are the obtained processing times for the bidirectional communications between Blender® and CODESYS®. The connection time between Blender® and CODESYS® is approximately 10 ms (Fig. 7). The processing time for reading data from input registers is 40 ms (Fig. 8). The processing time for writing data to holding

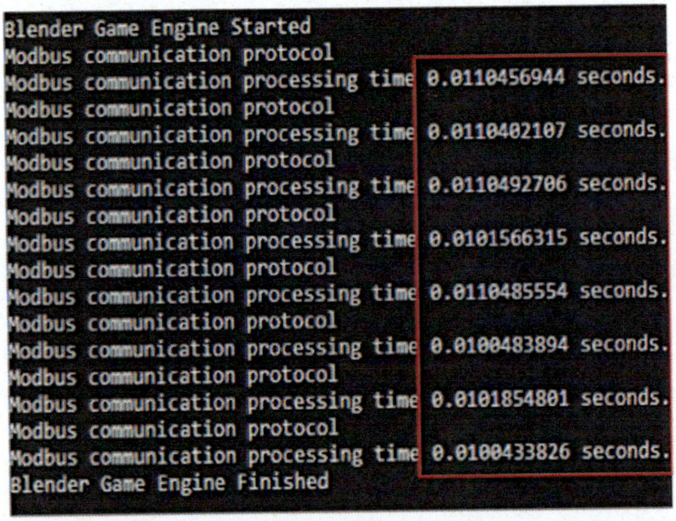

Fig. 7 Connection processing time

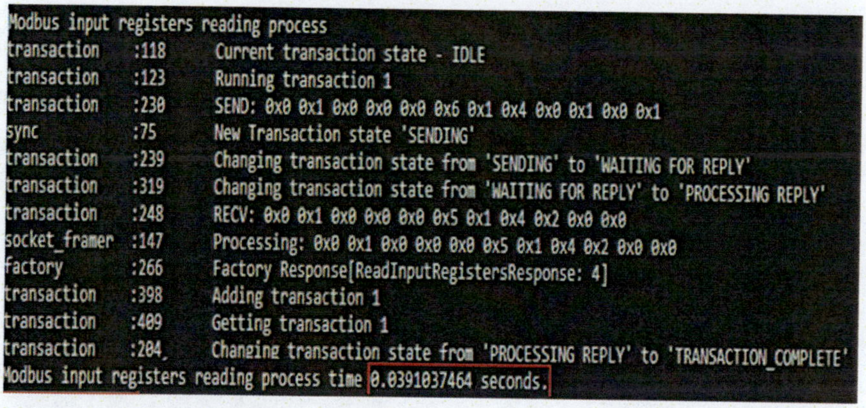

Fig. 8 Input registers reading processing time

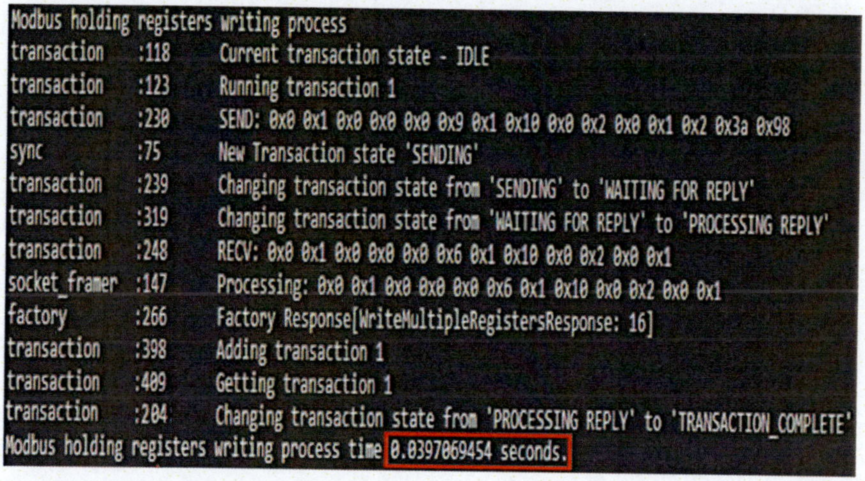

Fig. 9 Holding registers writing process time

registers is also 40 ms (Fig. 9). Therefore a sample time of 100 ms is enough for these communication operations.

5.3 *Functional Test of the Virtual Router Machine*

The virtual machine under operation is shown in Fig. 10, where there is an active interchange of signals from the virtual machine to the CODESYS® SoftPLC with the user designed control program. This figure shows the X, Y, and Z parameters that

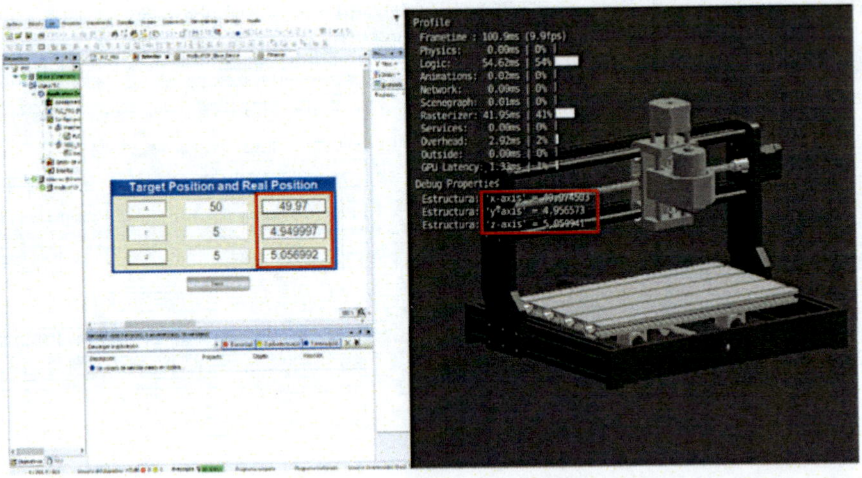

Fig. 10 Control program in CODESYS® (left) and screen shot of the virtual machine (right)

match in both the monitoring window of the SoftPLC and the animated positions of the virtual router machine at a specific time instant.

The Blender® software was a useful platform for animation of the 3D graphics of the machine. The user can develop the control program in a PLC, based on the inputs and outputs defined for interacting with the virtual machine. The virtual machine would then work for testing custom control strategies to perform specific tasks with the router. The user program should determine desired positions according to a trajectory generator, as well as the controller to produce the required manipulations for the actuators of the virtual machine.

6 Conclusions

Industry 4.0 implies a higher degree of automation and digitalization of production systems where the cyberphysical system is a key concept. Virtual machines are precursory computational models of digital twins and cyberphysical systems. Therefore, automation and design of control systems with open protocols such as Modbus are valuable skills that need to be developed both at universities and industries. Modern digital tools for online training and distance education should be pursuit.

The herein developed virtual machine incorporates desirable characteristics: (1) to be an open platform available to communicate with any brand of PLC that can be configured as a Modbus server device, (2) the communications interface signals are only the ones required by the actuators and sensors of the physical machine, and (3) neither modeling nor simulation functions need to be programmed in the PLC, and no control functions are available in the virtual machine. In this sense, the implemented systems result in a suitable virtual machine that is useful for training in PLC programming and for the development of important technical skills in industrial automation.

The proposed virtual router machine constitutes a learning platform that merges conventional automation with new digitalization capabilities. Further work on the presented virtual machine to turn it into a fully developed virtual twin is intended to achieve potential improvements in the operation of the real machine.

References

1. Dalenogare LS, Benitez GB, Ayala NF, Frank AG (2018) The expected contribution of Industry 4.0 technologies for industrial performance. Int J Prod Econ 204:383–394. https://doi.org/10.1016/j.ijpe.2018.08.019
2. Xu Y, Sun Y, Liu X, Zheng Y (2019) A digital-twin-assisted fault diagnosis using deep transfer learning. IEEE Access 7:19990–19999. https://doi.org/10.1109/ACCESS.2018.2890566
3. Orive D, Iriondo N, Burgos A, Saráchaga I, Álvarez ML, Marcos M (2019) Fault injection in Digital Twin as a means to test the response to process faults at virtual commissioning. In: 2019 24th IEEE international conference on emerging technologies and factory automation (ETFA), pp 1230–1234. https://doi.org/10.1109/ETFA.2019.8869334
4. Fitz T, Theiler M, Smarsly K (2019) A metamodel for cyber-physical systems. Adv Eng Inform 41:100930. https://doi.org/10.1016/j.aei.2019.100930
5. Senra-Simoes JD, Rodrigues-de-Seabra EA, Monteiro-Fernandes A (2017) Study, design and development of a Modbus master that evaluates the Modbus Communication between equipments. In: Silva-Gomes JF, Meguid SA (eds) Proceedings of the 7th international conference on mechanics and materials in design, Albufeira/Portugal, pp 1817–1818
6. Salamea HMT, Torres DDT, Cardenas PDU, Oñate CU (2020) Implementation of the Hamming code for the detection and correction of errors in a telerobotic system using an industrial communication protocol. In: 2020 IEEE ANDESCON, pp 1–6. https://doi.org/10.1109/ANDESCON50619.2020.9272049
7. Huang HB (2021) Communication between virtual emulation system and PLC MODBUS/TCP protocol. Accessed 03 May 2021. Available: https://core.ac.uk/reader/289247247
8. Trauer J, Schweigert-Recksiek S, Engel C, Spreitzer K, Zimmermann M (2020) What is a Digital Twin?—Definitions and Insights from an industrial case study in technical product development. Proc Des Soc Des Conf 1:757–766. https://doi.org/10.1017/dsd.2020.15
9. Bazaz SM, Lohtander M, Varis J (2019) 5-Dimensional definition for a manufacturing Digital Twin. Procedia Manuf 38:1705–1712. https://doi.org/10.1016/j.promfg.2020.01.107
10. Negri E, Fumagalli L, Macchi M (2017) A review of the roles of digital twin in cps-based production systems. Procedia Manuf 11:939–948. https://doi.org/10.1016/j.promfg.2017.07.198
11. Jones D, Snider C, Nassehi A, Yon J, Hicks B (2020) Characterising the Digital Twin: a systematic literature review. CIRP J Manuf Sci Technol 29:36–52. https://doi.org/10.1016/j.cirpj.2020.02.002
12. Tao F, Sui F, Liu A, Qi Q, Zhang M, Song B et al (2019) Digital twin-driven product design framework. Int J Prod Res 57(12):3935–3953. https://doi.org/10.1080/00207543.2018.1443229
13. Damiani L, Demartini M, Giribone P, Maggiani M, Revetria R (2018) Simulation and Digital Twin based design of a production line: a case study. In: Proceedings of the international multiconference of engineers and computer scientists, Hong Kong

14. Zhang H, Ma L, Sun J, Lin H, Thürer M (2019) Digital Twin in services and industrial product service systems: review and analysis. Procedia CIRP 83:57–60. https://doi.org/10.1016/j.procir.2019.02.131
15. Morales-Menendez R, Ramírez-Mendoza RA, Guevara AV (2019) Virtual/remote labs for automation teaching: a cost effective approach. IFAC-Pap 52(9):266–271. https://doi.org/10.1016/j.ifacol.2019.08.219
16. Alvarez J, Diaz G, Macias M (2019) Programming logical controllers using remote labs and virtual reality. In: 2019 IEEE international conference on engineering Veracruz (ICEV), Boca del Rio, Veracruz, Mexico, pp 1–4. https://doi.org/10.1109/ICEV.2019.8920533
17. Fox MFJ, Werth A, Hoehn JR, Lewandowski HJ (202) Teaching labs during a pandemic: Lessons from spring 2020 and an outlook for the future. ArXiv200701271 Phys. Accessed: 15 Mar 2021. Available: https://arxiv.org/abs/2007.01271

A Multispectral U-Net Framework for Crop-Weed Semantic Segmentation

Daniel Leyva Rosas, Uziel Grajeda Gonzalez, and Victor Gonzalez Huitron

Abstract The ability to visualize and understand the environment is essential for farming robots and precision agriculture. It allows them to act as the situation requires and take the necessary actions for a given scenario. The most recent state-of-the-art approaches use deep learning to generate models capable of crop classification, or detecting diseases and pests to reduce herbicides and pesticides, allowing a considerable reduction in pollution produced by the agricultural sector. In this work, we propose implementing a Fully-Convolutional Network (FCN) to perform semantic segmentation tasks. We use the U-Net architecture to visualize crops and weeds, initially proposed for use in medical imaging. For our implementation, we used the Sunflower dataset, which combines images in the Near-Infrared spectrum (NIR) and three channels images (RGB). The proposed FCN was evaluated employing Intersection over Union (IoU), with satisfactory results in semantic segmentation tasks.

Keywords Semantic segmentation · Deep learning · Precision agriculture

1 Introduction

Precision agriculture is a farming management strategy oriented at supporting decision-making based on the observation, measurement, analysis, and processing of data that allows us to understand the current state of the soil or crops and their variations. The main objective of precision agriculture is to reduce the environmental impact of pollution produced by the agricultural sector by reducing pesticides, herbicides, and water, among others [10]. In this matter, the use of drones and farm-

D. L. Rosas (✉) · U. G. Gonzalez
Tecnologico Nacional de México Campus Culiacén, Juan de Dios Batiz No. 310 Pte., Col. Guadalupe, 80220 Culiacan Rosales, Mexico
e-mail: Daniel_leyva@itculiacan.edu.mx

V. G. Huitron
CONACyT-Tecnológico Nacional de México/Instituto Tecnológico de Culiacán, Sinaloa, Mexico

K. L. Flores Rodríguez et al. (eds.), *Recent Trends in Sustainable Engineering*, Lecture Notes in Networks and Systems 297,
https://doi.org/10.1007/978-3-030-82064-0_2

Fig. 1 Farming robots for data adquisition, **a** Bonirob system used in [11] for data adquisition, **b** Bosch Bonirob Farming robot used in [3] for data adquisition

ing robots in conjunction with perception systems are beneficial (Fig. 1), as they are capable of achieving automation, classification, detection, and similar tasks [3].

Some novel solutions, which use perception systems based on advanced machine learning techniques, can offer plant-level weed control treatments through selective spraying systems or mechanical actuators. This task, however, requires classification systems capable of collecting data and images on the field, analyzing them, and generating information that allows us to understand the environment and ease decision-making processes [11].

Models based on Convolutional Neural Networks (CNN) have shown a straightforward utility in solving image processing and computer vision problems. They achieve a deep understanding of the analyzed images patterns and have been widely used for classification tasks in plants [2, 9, 17].

The utility of traditional CNN's has been demonstrated for image classification tasks [6, 18], object detection [7, 14, 20] and image segmentation [1, 12, 16]; [8] shows particularly, applications for semantic segmentations. Altogether, Fully Convolutional Networks are especially useful compared to CNN.

In Sect. 2 we see how CNN's have evolved and have been used for increasingly specialized tasks, we discuss how state-of-the-art models address semantic segmentation problems, and the way the use of multispectral images has become generalized in the area. In Sect. 3 we explain in a general way the U-Net architecture and the Sunflower Dataset, as well as how they are integrated to achieve good performance. In Sect. 4 we explain in detail the data and parameters used for training the model and the quality metrics obtained. Also in this section, we show some of the results obtained with our model. And finally, in Sect. 5, we express our conclusions.

2 Related Work

In recent years, multiple models based on supervised learning have been proposed addressing the crop classification task. A feasible hardware implementation is mandatory due to the agroindustry requirements that adopt in short time innovative solutions [9, 17]. Various efforts have been made to develop algorithms that approach the segmentation task. Notably, the design of models capable of crop disease detection through the use of convolutional neural networks have achieved outstanding results [4, 9, 13, 17], which facilitates the possibility of implementing more competent and capable precision agriculture systems through farming robots and drones.

The most recent, novel implementations employs models based on semantic segmentation [3, 11, 12, 16], as it allows to specifically isolate the factor of interest, allowing for a more accessible, cleaner visualization. The use of algorithms based on CNN have been a hot topic research in computer vision tasks [6], where the emergence of algorithms capable of large-scale image processing [18] allowed the development of increasingly specialized models.

In [1] it can be noted how a topology identical to the convolutional layers presented in [18] is used successfully, resulting in the already known SegNet.

Further specialization is achieved by Sa et al. [16]; proposing a model for crop/weed semantic segmentation, based on the SegNet encoder-decoder architecture, called weedNet. Sa et al. [16] also suggests using multispectral images in its solution, stating that these provide the possibility of working with indices based on radiance ratios that are more robust under varying lighting conditions, making its use suitable for farming robots.

Moreover, the use of multispectral images in models oriented to agricultural applications, can be observed in datasets such as sugar beet [2] and Sunflower [3]. References [9, 13, 16, 19] also make use of images with extra channels additional to traditional RGB in order to improve results by increasing the number of features to process.

3 Proposed Method

We approach the crop-weed detection problem by using semantic segmentation techniques based on Fully convolutional networks and multispectral images in this work. The main workflows are as follows: first, a Near-infrared (NIR) image spectrum is combined with a three-channel (RGB) image; thus, a four-channel array is employed as input for the CNN model. Secondly, an encoding-decoding CNN processes the data array in order to obtain a 4-channels array, consisting of a segmented mask image and three channels for RGB segmented imaging. Finally, is obtained pixel-level classification for crop, weed, and soil from these results. A workflow is presented in Fig. 2.

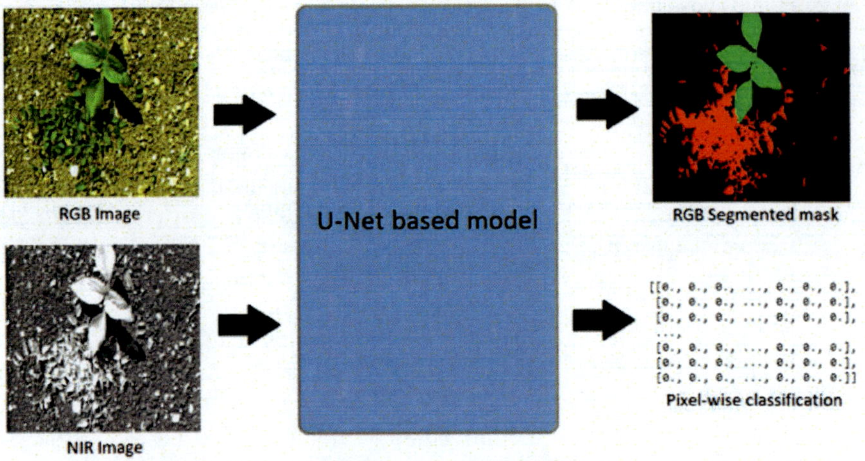

Fig. 2 Representation of the proposed flow for RGB and NIR images as inputs, and RGB segmented mask image and pixel-level classification array as outputs

Table 1 Hyperparameters used for multispectral U-Net models training

Hyperparameter	Value
Optimization algorithm	Adam
Epochs	300
Batch size	8–16
Loss function	Binary crossentropy
Learning rate	0.0001
DropOut value	0.05
Num filters	16

For our proposal model, an architecture based on U-Net [15] is adapted for its use with multispectral images. It has shown to be capable of generating models of notable performance training on relatively small datasets. As training data, the Sunflower dataset was used for semantic crop-weed segmentation [3]. In Table 1 we show the hyperparameters used on the training of our models for both experiments.

This implementation is based on the tensorflow, keras and sklearn libraries for the creation and training of the model, as well as numpy and skimage for the handling of images and arrays. The model training was performed on the Google Colab platform in conjunction with Google drive to save and access the datasets and models.

3.1 Architecture

The U-Net architecture is based on an encoder-decoder model commonly employed in deep learning for segmentation tasks. The convolutional layers perform feature extraction. Then, upsampling layers estimate an output image from the extracted features, as have been implemented in previous works such as segnet [1], or weednet [16].

U-Net also combines the features maps extracted in each downsampling stage with its upsampling counterpart in a similar fashion that resembles the architecture proposed on resNet [5], which allows the network to deepen on features extraction while preserving more general context information. As stated in Fig. 3. Each blue box corresponds to a multi-channel feature map. White boxes represent feature maps copied from previous layers and each arrow represents a different action denoted in the image bottom-right corner.

3.2 Dataset

As stated in [3] the creation of a semi-artificial dataset for crop-weed semantic segmentation is possible with a novel approach that propounds modifying only the areas of interest in the image, which correspond to plants and crops, and leaving

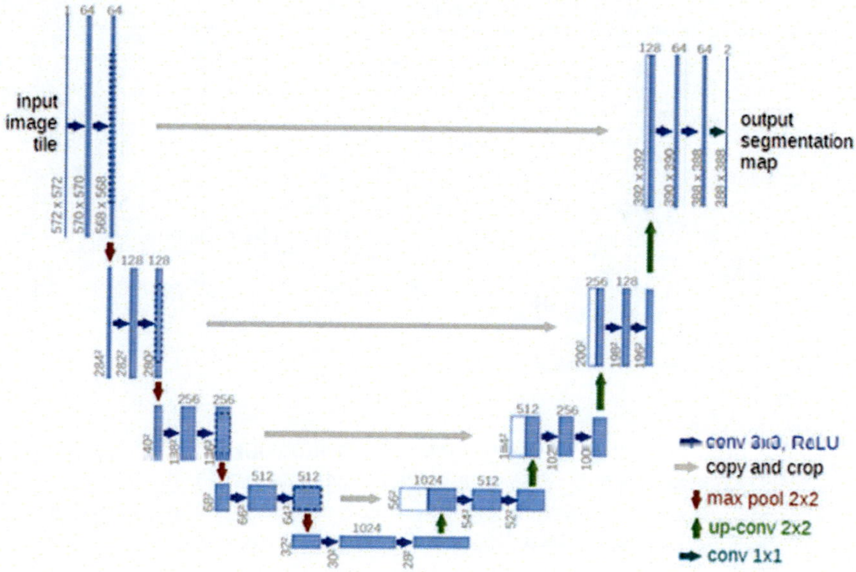

Fig. 3 U-Net architecture topology and layers, as shown in [15]

Fig. 4 An example of an original image and the synthetic one generated using the cGAN. image extrated from [3]

Table 2 Pixel-wise segmentation performance for sunflower dataset, trained on RGB + NIR input

Dataset	Model	Mean IoU	Soil IoU	Crop IoU	Weed IoU
Syntetic crop	Bonnet	0.83	0.99	0.84	0.66
	U-Net	0.701	0.99	0.69	0.41
	U-Net-resNet	0.704	0.99	0.65	0.47
	Bonnet	0.80	0.99	0.78	0.62
Original	U-Net	0.64	0.99	0.54	0.38
	U-Net-resNet	0.66	0.99	0.60	0.43
	Bonnet	**0.86**	**0.99**	**0.88**	**0.69**
Mixed	U-Net	0.51	0.99	0.68	0.40
	U-Net-resNet	0.53	0.99	0.70	0.48

Extracted from [3]
Best results are shown in bold

the rest of the original image intact. A conditional generative adversarial network (cGAN) was used to create semi-artificial data instances, replacing the original crops and weeds with synthetic images. In Fig. 4, an example is shown. The methodology was used for both RGB and NIR dataset creation.

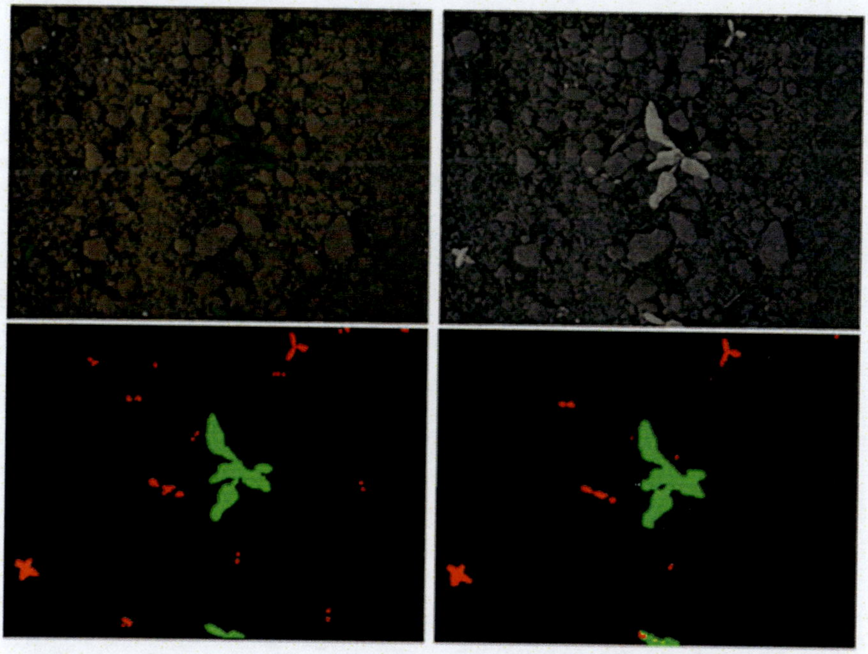

Fig. 5 Extracted from [3]. Top row: from left to right, synthetic RGB and synthetic NIR samples, respectively. Bottom row: from left to right, the pixel-wise ground-truth and the result obtained by using a semantic segmentation deep neural network, respectively

The experiments carried out in [3] show how the use of mixed datasets (original - enhanced) generates better results compared to the use of completely artificial or original datasets as shown in Table 2 extracted from the original article.

The dataset contains a three-channel RGB image, as well as a one-channel image in the NIR spectrum. For ground-truth and comparison purposes, an RGB segmented mask and a one-channel class mask is provided. In Fig. 5, a sample data is shown.

As is often the case on datasets designed for segmentation tasks, this dataset faces the problem of manual labeling, so it does not contain a substantial amount of data. In total, the Sunflower dataset has 500 examples distributed in 3 subsets; each stage includes a combination of original and semi-artificial images.

4 Results

For our proposal, two of the three parts of the Sunflower dataset (jesi_05_12 and jesi_05_18) were taken as a training set, using both RGB and NIR images rescaled to a resolution of 512 * 512 pixels for experiment 1, and 704 * 704 pixels for experiment 2. The third part of the dataset (jesi_06_13) was used as validation data.

Table 3 Objective results by means of Jaccard index (IoU) values for the Sunflower dataset

Model	Crop IoU	Weed IoU	Soil IoU	Mean IoU
U-Net (Ours, 704 * 704)	**0.90**	**0.76**	0.86	0.84
U-Net (Ours)	0.88	0.75	0.83	0.82
U-Net [3]	0.68	0.40	0.99	0.69
Bonnet [3]	0.88	0.69	**0.99**	**0.86**
U-net-resnet [3]	0.70	0.48	0.99	0.72

Best results are shown in bold

Table 3 shows the values obtained by measuring the Jaccard index as seen in Eq. (1), also known as intersection over union (IoU), for each of our models compared to the results reported in [3].

$$D_{Jaccard}\ (A, B) = \frac{|A \cap B|}{|A \cup B|} \tag{1}$$

The IoU value for each class is the arithmetic average of the value obtained per class for the total images in each subset. To obtain the Mean IoU value, the arithmetic average of the IoU value for each of the three classes was calculated.

In our training and testing, we were limited by hardware capacity, which restrained the use of the complete dataset for training, as well as using larger image sizes. Despite this, the results obtained are helpful for implementations oriented to precision agriculture, as Table 3 states. Figure 6 shows some examples of the results obtained in this work.

The performance of the network is generally superior to the results obtained in the experiments carried out in [3] using the same dataset. It is important to note that in [3] the complete Sunflower dataset (500 images) with an image size of 512 * 512 pixels was used. In comparison, for our implementation, only two subsets (318 images) were used with a different size on each experiment (512 * 512 pixels for experiment 1 and 704 * 704 pixels for experiment 2).

In Table 3, we can see how the IoU values measured for our 512 * 512 pixel model are higher for the crop and weed classes compared to the U-Net, U-Net-resNet and Bonnet models presented in [3], while obtaining lower IoU values for the soil class which results in a lower Mean IoU value. Despite this, we consider that the crop and weed classes are more significant for precision agriculture uses, and a slightly higher margin of error in the soil class does not represent a reduction in the model functionality.

It can be seen that despite having a considerably smaller training set, our implementation obtains significantly higher values than those obtained by [3] for U-Net and U-Net-resnet, as well as slightly higher values than those obtained with Bonnet. Finally, Table 2 also compares our 704 * 704 pixel model, showing better results for crop and weed classes compared to our 512 * 512 model and the three models

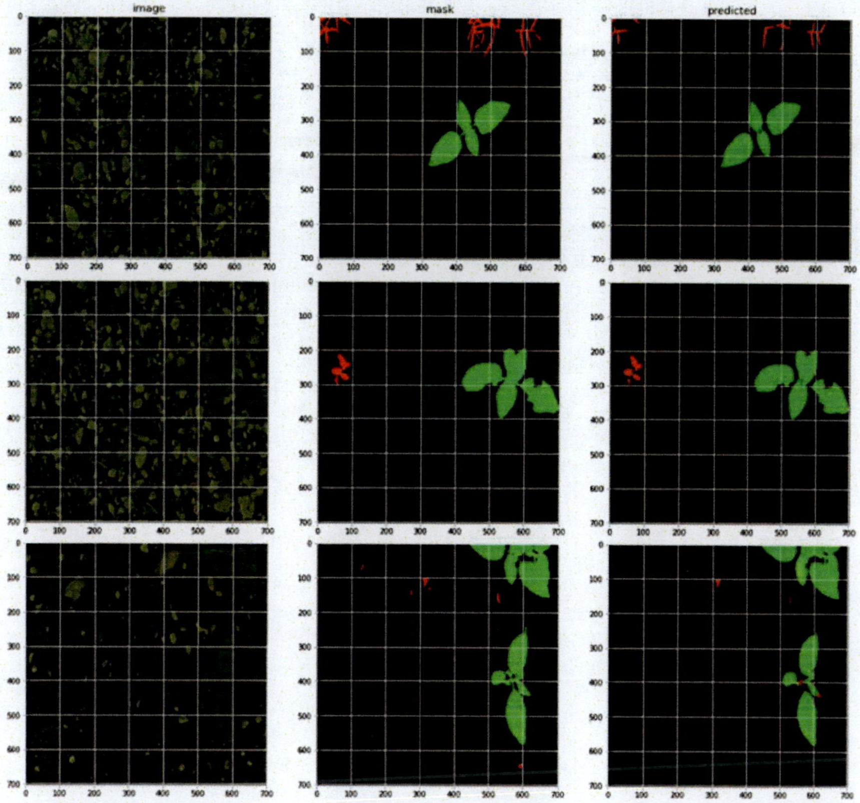

Fig. 6 From left to right: original input RGB+NIR image, ground truth segmented RGB mask, segmented RGB mask obtained with our model

presented in [3], which allows us to infer that larger image sizes should obtain better results.

5 Conclusions

The U-net architecture is a handy tool for tasks oriented to semantic segmentation in non-medical areas. Since it is possible to obtain reliable results with a surprisingly small amount of training data, its use can significantly help precision agriculture tasks.

For the relatively large training datasets scenario, U-net might not be the best fit. There are alternatives such as Bonnet or segNet that can obtain better results, translating into shorter processing times if the required hardware is available. Still, these architectures usually need more data to get a better performance.

For semantic segmentation tasks with a limited training dataset, a U-net implementation with relatively large input image sizes is an excellent choice.

References

1. Badrinarayanan V, Kendall A, Cipolla R (2016) SegNet: a deep convolutional encoder-decoder architecture for image segmentation. arXiv:151100561 [cs]
2. Chebrolu N, Lottes P, Schaefer A et al (2017) Agricultural robot dataset for plant classification, localization and mapping on sugar beet fields. Int J Robot Res 36:1045–1052. https://doi.org/10.1177/0278364917720510
3. Fawakherji M, Potena C, Pretto A et al (2020) Multi-spectral image synthesis for crop/weed segmentation in precision farming. arXiv:200905750 [cs]
4. Fuentes A, Yoon S, Kim S, Park D (2017) A robust deep-learning-based detector for real-time tomato plant diseases and pests recognition. Sensors 17:2022. https://doi.org/10.3390/s17092022
5. He K, Zhang X, Ren S, Sun J (2015) Deep residual learning for image recognition. arXiv:151203385 [cs]
6. Lecun Y, Bottou L, Bengio Y, Haffner P (1998) Gradient-based learning applied to document recognition. Proc IEEE 86:2278–2324. https://doi.org/10.1109/5.726791
7. Liu W, Anguelov D, Erhan D et al (2016) SSD: single shot MultiBox detector. arXiv:151202325 [cs] 9905:21-37. DOIurl10.1007/978-3-319-46448-0_2
8. Long J, Shelhamer E, Darrell T (2015) Fully convolutional networks for semantic segmentation. arXiv:14114038 [cs]
9. Lowe A, Harrison N, French AP (2017) Hyperspectral image analysis techniques for the detection and classification of the early onset of plant disease and stress. Plant Methods 13. https://doi.org/10.1186/s13007-017-0233-z
10. McBratney A, Whelan B, Ancev T, Bouma J (2005) Future directions of precision agriculture. Precis Agric 6:7–23. https://doi.org/10.1007/s11119-005-0681-8
11. Milioto A, Lottes P, Stachniss C (2018) Real-time semantic segmentation of crop and weed for precision agriculture robots leveraging background knowledge in CNNs. In: 2018 IEEE International conference on robotics and automation (ICRA). IEEE
12. Milioto A, Stachniss C (2019) Bonnet: an open-source training and deployment framework for semantic segmentation in robotics using CNNs. In: 2019 International conference on robotics and automation (ICRA). IEEE
13. Pourazar H, Samadzadegan F, Dadrass Javan F (2019) Aerial multispectral imagery for plant disease detection: radiometric calibration necessity assessment. Eur J Remote Sens 52:17–31. https://doi.org/10.1080/22797254.2019.1642143
14. Redmon J, Divvala S, Girshick R, Farhadi A (2016) You only look once: unified, real-time object detection. arXiv:150602640 [cs]
15. Ronneberger O, Fischer P, Brox T (2015) U-Net: convolutional networks for biomedical image segmentation. arXiv:150504597 [cs]
16. Sa I, Chen Z, Popovic M et al (2018) weedNet: dense semantic weed classification using multispectral images and MAV for smart farming. IEEE Robot Autom Lett 3:588–595. https://doi.org/10.1109/lra.2017.2774979
17. Saleem Potgieter, Arif Mahmood (2019) Plant disease detection and classification by deep learning. Plants 8:468. https://doi.org/10.3390/plants8110468
18. Simonyan K, Zisserman A (2015) Very deep convolutional networks for large-scale image recognition. arXiv:14091556 [cs]
19. Sosa-Herrera JA, Vallejo-Pérez MR, Álvarez-Jarquín N et al (2019) Geographic object-based analysis of airborne multispectral images for health assessment of *Capsicum annuum L.* crops. Sensors 19:4817. https://doi.org/10.3390/s19214817
20. Wong A, Shafiee MJ, Li F, Chwyl B (2018) Tiny SSD: a tiny single-shot detection deep convolutional neural network for real-time embedded object detection. arXiv:180206488 [cs]

Design and Simulation of a Neural Controller for MIMO Systems

Roberto S. Velazquez-Gonzalez, Julio C. Sosa-Savedra, and Agustin Barrera-Navarro

Abstract Classical control theory is widely used to regulate Simple Input, Simple Output (SISO) systems. However, when there are multiple inputs and multiple outputs, the complexity to control the systems increases significantly. Intelligent control theories such as artificial neural networks have proven to be an efficient alternative to implement complex systems. In addition, it is not necessary to obtain a mathematical model of the system. This paper shows the application of artificial neural networks as control elements in MIMO systems (Multiple Input, Multiple Outputs). The performance of the controller is shown in a simulation of a combined system, in which two first-order plants are interconnected. The experiments were carried out in Matlab software.

Keywords Neural controller · Backpropagation algorithm · Artificial neural network · MIMO system

1 Introduction

In many applications, it is very common to control multiple processes, whose inputs affect directly the response of the outputs, these kinds of systems are commonly called MIMO systems (Multiple Input, Multiple Output). There are many approaches to control MIMO systems, for instance, it is a common industrial practice to reduce a multivariable control problem to SISO (Simple Input, Simple output) control approach. In this sense, Ciprian [4] proposes some classical techniques to control MIMO systems based on decentralized control, static decoupling procedures, and a combined feedforward - feedback structure.

R. S. Velazquez-Gonzalez (✉) · J. C. Sosa-Savedra
Instituto Politécnico Nacional, CICATA -Querétaro, Querétaro, Mexico
e-mail: jcsosa@ipn.mx

A. Barrera-Navarro
Tecnológico Nacional de México, Querétaro, Mexico
e-mail: abarrera@mail.itq.edu.mx

K. L. Flores Rodríguez et al. (eds.), *Recent Trends in Sustainable Engineering*,
Lecture Notes in Networks and Systems 297,
https://doi.org/10.1007/978-3-030-82064-0_3

On the other hand, classical control techniques, such as PI, PID, among others, require a mathematical model for the design of the process controller, otherwise, it is difficult to achieve a satisfactory result. A model is an abstract representation of the process to be controlled. In many cases, it is very difficult to model the system to be controlled, so non-parametric methods are used for this purpose [13].

Other approaches to control MIMO systems include non-classical control techniques, like fuzzy logic, Expert Systems, Genetic Algorithms, Artificial Neural Networks (ANNs), and so on. Some contributions in this field are described in [9, 11, 15].

According to the literature, the complexity for the control of MIMO systems increases when there are coupled systems, that is, when the behavior of one variable affects the state of another, such is the case of the regulation of pH and electrical conductivity in nutrient solutions of hydroponic crops.

In this work, a controller based on ANN is proposed for MIMO systems, with two first-order systems and using the back propagation of the error algorithm. The results obtained through simulation show that despite the complexity of interconnecting two systems, the neural network is able to maintain the values of the variables within the specified ranges.

The present document is organized as follows: in Sect. 2, a general theoretical framework of the neural networks applied in control theory is shown. Section 3 presents the mathematical foundations for the proposed model of the controller. In Sect. 4 the results of some simulations are presented and the discussion of the performance of the controller. In Sect. 5 the conclusion about this work is presented. Finally, in Sect. 6 we present future work to complement this research.

2 Neuro-adjustable Controller

ANNs are defined as non-linear mapping systems whose structure is based on principles observed in the nervous systems of humans and animals [12]. The processing units are called neurons. Each unit receives inputs from other nodes and generates a simple scalar output that depends on the available local information, stored internally, or that comes through the weights of the connections [7]. Within their wide range of applications, they are ideal for controlling systems that for some specific reason are difficult to model [1, 16]. Another important feature of the neural networks used in control applications is their easy adaptation to the environment and the learning capacity, simplifying the task of controller design [3, 5, 10].

A general outline of an ANN is shown in Fig. 1, where $i_1, i_2, \ldots, i_n$ are the inputs of the neural network, and it could represent the current error of the system, the error of a past epoch, and the error of n past epochs respectively. On the other hand, input and hidden layer weights represent the adjustable weights in each learning iteration, these are multiplied by the inputs and added together so that the result is entered into an activation function, which is usually a sigmoid function ϕ and the output of that function is the output of the neuronal controller.

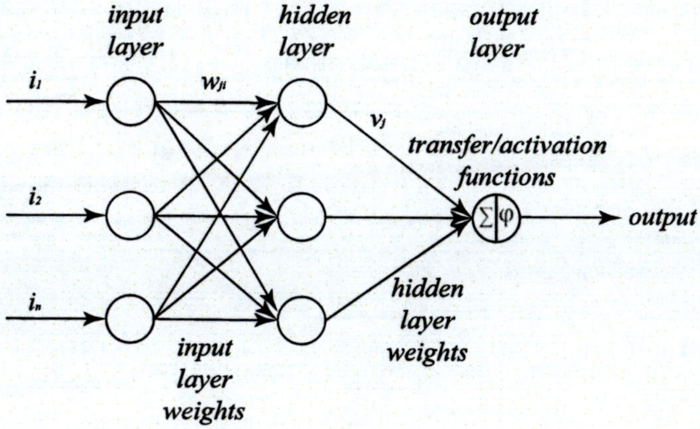

Fig. 1 General scheme of an ANN

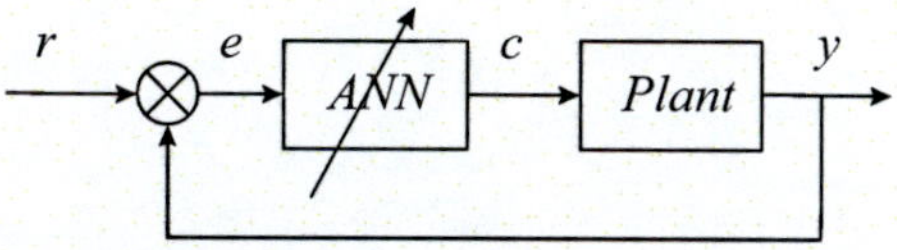

Fig. 2 General scheme of the neuro-adjustable controller

Figure 2 shows a self-adjusting regulator inspired in Cui and Shin controller [6], but with the particular characteristic that the previous training is replaced by a permanent adjustment of the weight coefficients based on the adjustment error.

Where r is the desired value or Set Point, e, is the error, C, the control output, and y, the response of the system to be controlled. The neuronal network that acts as a regulator is a perceptron of 3 layers of neurons (one hidden layer), which weight coefficients are adjusted using the back propagation of the error algorithm [2, 14]. In this case, however, instead of the mains output error (1), the process output error (2) is used to adjust the coefficients of each weight of the neural regulator.

$$e_u(t) = u_d(t) - u(t) \tag{1}$$

$$e_y(t) = y_r(t) - y(t) \tag{2}$$

Figure 3 shows the structure of the network that acts as a controller. In the input layer, the current value and some previous values of the control error are entered, denoted in (3). The output of the jth neuron of the hidden layer is calculated as it is shown in (4).

$$x(t) = [e_y(t), e_y(t-1), \ldots, e_y(t-n)] \tag{3}$$

Fig. 3 Internal structure of neuro-adjustable controller

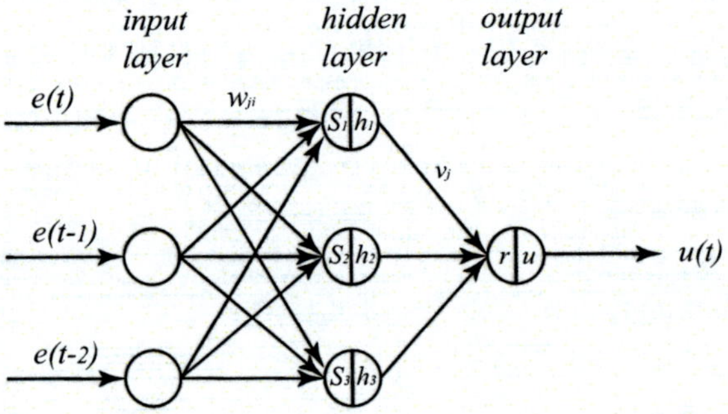

$$h_j = \frac{1}{1 + e^{-S_j}} \tag{4}$$

where

$$S_j = \sum_{i=1}^{3} w_{ji} x_i \tag{5}$$

and

$$r = \sum_{j=1}^{3} v_j h_j \tag{6}$$

On the other hand, the output of the neural network, which acts how control signal for the system is:

$$u = \frac{1}{1 + e^{-r}} \tag{7}$$

Likewise, the error minimization criterion is defined as:

$$E(t) = \frac{1}{2} \sum_{k=1}^{t} e_y(k)^2 \tag{8}$$

The minimization procedure consists of moving in the direction of the negative of the gradient of the $E(t)$ function with respect to the weight coefficients v_j and w_{ji}. The $E(t)$ gradient is a multidimensional vector which components are the partial derivatives shown in (9).

$$\nabla E(t) = \begin{bmatrix} \frac{\partial E(t)}{\partial v_j} \\ \frac{\partial E(t)}{\partial w_{ji}} \end{bmatrix} \tag{9}$$

Using the backpropagation error, the partial derivatives are defined with respect to the weight coefficients of the output neuron as:

$$\frac{\partial E(t)}{\partial v_j} = \delta^1 h_j \frac{\partial e_y}{\partial e_u} \tag{10}$$

where:

$$\delta^1 = e_y u(t)(1 - u(t)) \tag{11}$$

In (10), the partial derivative $(\partial e_y)/(\partial e_u)$ can be interpreted as the equivalent gain of the process and according to Cui and Shin [6] can be replaced by a constant coefficient of +1 or −1.

The partial derivative of the $E(t)$ function with respect to the w_{ji} coefficients is obtained by applying the chain rule again (12).

$$\frac{\partial E(t)}{\partial w_{ji}} = \frac{\partial E(t)}{\partial e_y}\frac{\partial e_y}{\partial e_u}\frac{\partial e_u}{\partial e_u(t)}\frac{\partial e_u(t)}{\partial r}\frac{\partial r}{\partial h_j}\frac{\partial h_j}{\partial s_j}\frac{\partial s_j}{\partial w_{ji}} \tag{12}$$

3 Design

3.1 Development of the Neural Controller

The proposal of the system to be controlled, as well as the controller, are shown in Fig. 4. The dynamics consists of two first-order systems that interact, since the output of the first block affects the operation of the second block. In this architecture, current and previous errors are used as input signals to the neural networks for the two blocks, on the other hand, for the output neurons, errors from the opposing blocks are used.

The outputs of the u_1 and u_2 neurons are given by (13) and (14).

$$u_1 = \frac{1}{1 + e^{-r_1}} \tag{13}$$

$$u_2 = \frac{1}{1 + e^{-r_2}} \tag{14}$$

where:

$$r_1 = \sum_{j=1}^{4} v_{j1} h_j \tag{15}$$

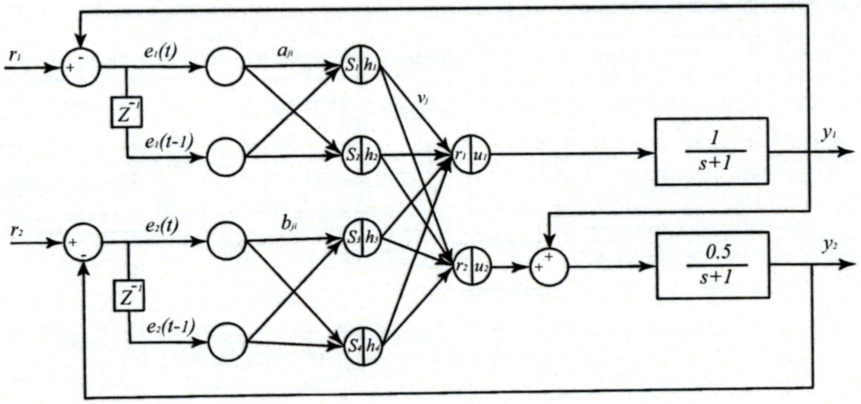

Fig. 4 Proposal of the neuro-adjustable controller

and:

$$r_2 = \sum_{j=1}^{4} v_{j2} h_j \tag{16}$$

The outputs of the hidden layer are defined by (17)

$$h_i = \frac{1}{1 + e^{-s_i}} \Big|_{i=1}^{4} \tag{17}$$

And the transfer functions for the hidden layer are given by (18)–(21):

$$s_1 = e_1(t)a_{11} + e_1(t-1)a_{21} \tag{18}$$

$$s_2 = e_1(t)a_{12} + e_1(t-1)a_{22} \tag{19}$$

$$s_3 = e_2(t)b_{11} + e_2(t-1)b_{21} \tag{20}$$

$$s_4 = e_2(t)b_{12} + e_2(t-1)b_{22} \tag{21}$$

Using the backpropagation algorithm, the optimization equations for the weights of the input and hidden layers are obtained by (22), (23) and (24).

$$v_{ji} = v_{ji}(t-1) + \eta e_i(t) u_i (1 - u_i) h_j \tag{22}$$

$$a_{ji} = a_{ji}(t-1) + \eta e_1(t)u_1(1-u_1)v_{ji}h_j(1-h_j)e_1(t-1) \tag{23}$$

$$b_{ji} = b_{ji}(t-1) + \eta e_2(t)u_2(1-u_2)v_{ji}h_j(1-h_j)e_2(t-1) \tag{24}$$

where η, is considered a learning factor of the neural network, and varies in a range from 0 to 1.

3.2 Algorithm

Algorithm 1 shows the logic implemented in order to obtain the optimization equations for the weights.

Algorithm 1: Neural Weights Optimization

```
input  : Two set-points (yr1, yr2), Number of iterations (n)
output: Two responses (y1, y2)
weights initialization;
set points initialization;
it ⟵ 0;
while it ≤ n do
    error = reference - output;
    compute neuron outputs;
    compute new weights;
    compute control signal;
    it++;
end
plot results;
```

4 Results

4.1 Output Responses

To verify the operation of the controller, a simulation in Matlab® software was implemented. The recursive equations for plant 1 and plant 2 were obtained using the z-transformation method in order to digitize the transfer functions [8].

The first experiment corresponds to setting the setpoints for the systems at a fixed value. Figure 5 shows the response for the two systems with a Set-Point of 0.65 (SP1) and 0.35 (SP2) units for systems 1 and 2 respectively, which can represent

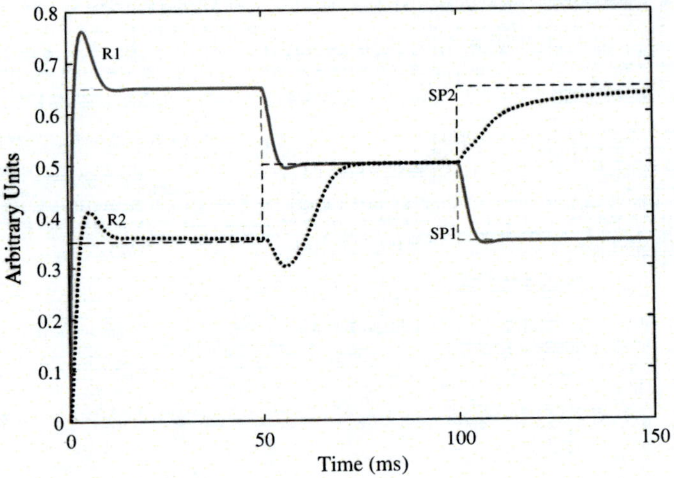

Fig. 5 Profile of the responses of the multivariable system

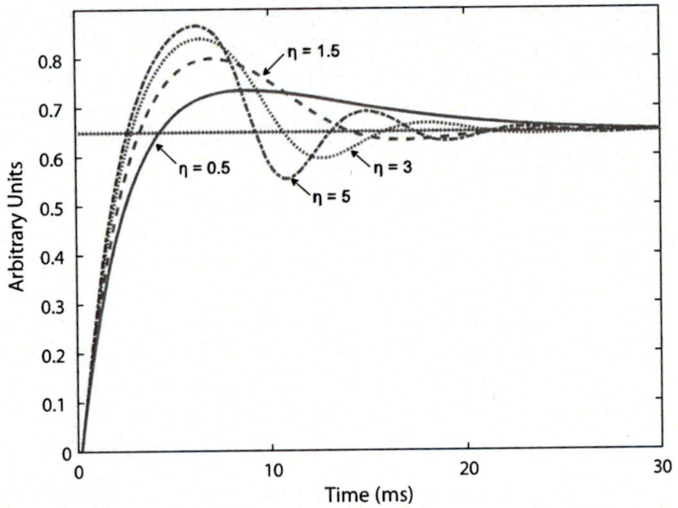

Fig. 6 Behaviour of the system 1 for different values of η

temperature, pressure, level, or any other variable. The references were changed every 50 ms. For this experiment, the neural network was not previously trained.

In the second experiment, the training factor was changed for system 1. Figure 6 shows the overshoot and the oscillations before the establishment time for five different values of η. For eta values greater than 0.5, the performance of the controller is not significantly better, in fact, oscillations appear and this makes the system unstable.

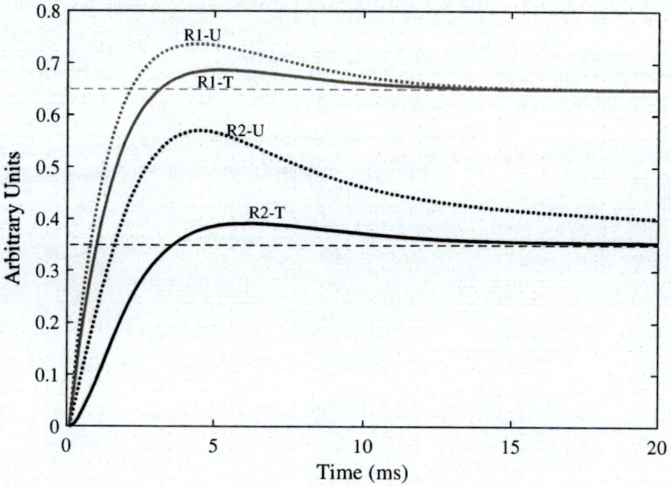

Fig. 7 Profile of the system outputs before and after of neural network training. R1-U and R2-U show the performance of output variables using random weights at the beginning of the simulation. On the other hand, R1-T and R2-T show the profile of output variables using the weights after a training process

In the third experiment, random weights were used at the start of the simulation, and after the output variables reached their references, the weights obtained were used to start the simulation again. Figure 7 shows the overshoot, which is reduced considerably, in addition, the establishment time is also reduced.

In order to evaluate the performance of the controller, additional experiments were carried out. The experiments consisted of changing the learning factor (η) during the training of the neural controller.

Once the output variable is established in their corresponding reference (R1 or R2), two parameters were evaluated: the overshoot (OS) and the time of establishment (TE) for each output response. The results are shown in Table 1.

The best performance is found when η takes values between 0.5 and 1, since the overshoot is established below 20% and the establishment time is less, in addition, there are no significant oscillations.

4.2 *Weights Profile*

The weights profile on the time for the input layer is shown in Fig. 8, the optimization equations modify the weights of each layer according to the error of the output. Once the error is minimized, the weights are stabilized in a constant value. Both

Table 1 Evaluation of controller performance

η	OS R1 (%)	TE R1 (ms)	OS R2 (%)	TE R2 (ms)
0.5	11	32	12	35
0.7	14	30	15	35
0.9	16	29	18	28
1	17	25	19	22
1.5	20	22	23	20
3	28	60	29	28
5	40	68	32	32

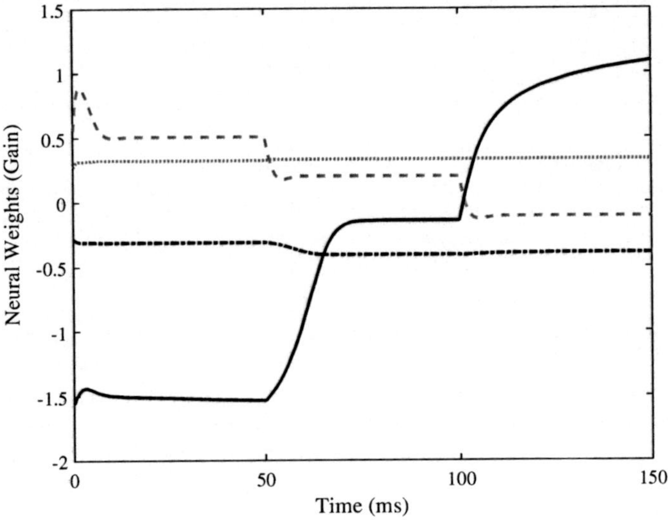

Fig. 8 Profile of weights during the training process

weights, input and hidden layers converge in a constant value in approximately 50 milliseconds. This time can be optimized by modifying the learning factor, however, the overshoot can increase.

5 Conclusions

In the present research work, a specific controller design based on ANN was presented for a multivariable system at the simulation level, and although the systems presented are first order, the interaction between both complicates the control of the process through conventional methods. Comparing this work with other classical control proposals, we can find some advantages: a mathematical model of the system to be controlled is not necessary to implement the controller. In addition, there are few parameters that need to be adjusted to improve the controller performance, for example, the learning factor. In summary, neural controllers represent a good option for the control of MIMO systems with a certain degree of complexity, otherwise, more sophisticated and difficult to implement control algorithms would be necessary.

6 Future Work

We recognize that to fully evaluate the performance of the proposed controller it is necessary to subject it to more realistic conditions, e.g., systems with delay and second or higher order systems. Also, it is necessary to analyze the stability of the system, however, this task is left for further work.

References

1. Abdallah C, Dawson D, Dorato P, Jamshidi M (1991) Survey of robust control for rigid robots. Control Syst 11:24–30 (03 1991). https://doi.org/10.1109/37.67672
2. Albarakati N, Kecman V (2013) Fast neural network algorithm for solving classification tasks: batch error back-propagation algorithm. In: 2013 Proceedings of IEEE Southeastcon. pp 1–8
3. Chen PCY, Mills JK (1997) Neural network generalization and system sensitivity in feedback control systems. In: Proceedings of 12th IEEE International symposium on intelligent control. pp 233–238
4. Ciprian L, Dumitru P, Andreea U, Catalin D (2008) Solutions for nonlinear multivariable processes control. WSEAS Trans Sys Ctrl 3(6):597–606
5. Cui X, Shin KG (1991) Design of an industrial process controller using neural networks. In: 1991 American control conference, pp 508–513
6. Cui X, Shin KG (1993) Direct control and coordination using neural networks. IEEE Trans Syst Man Cybern 23(3):686–697
7. Handelman DA, Lane SH, Gelfand JJ (1990) Integrating neural networks and knowledge-based systems for intelligent robotic control. IEEE Control Systems Magazine 10(3):77–87
8. Helm HA (1959) The z transformation. The Bell System Technical Journal 38(1):177–196
9. Hidalgo D, Bacca B, Caicedo Bravo E (2016) Control strategy based on swarms algorithms to cooperative payload transport using a non-holonomic mobile robots group. IEEE Latin America Trans 14:445–456 (02 2016). https://doi.org/10.1109/TLA.2016.7437178
10. Hunt K, Sbarbaro D, Zbikowski R, Gawthrop P (1992) Neural networks for control systems—a survey. Automatica 28(6):1083–1112
11. Lin X, Wu C, Chen B (2019) Robust h_∞ adaptive fuzzy tracking control for MIMO nonlinear stochastic Poisson jump diffusion systems. IEEE Trans Cybern 49(8):3116–3130
12. Ponce P (2015) Inteligencia artificial con aplicaciones a la ingeniera (05 2015)
13. Roeva O, Slavov T (2014) PID controller tuning based on metaheuristic algorithms for bioprocess control. Biotechnol Biotechnol Equip 26:3267–3277 (04 2014). https://doi.org/10.5504/BBEQ.2012.0065
14. Sagar V, Kumar K (2015) A symmetric key cryptography using genetic algorithm and error back propagation neural network. In: 2015 2nd International conference on computing for sustainable global development (INDIACom). pp 1386–1391

15. Xin LP, Yu B, Zhao L, Yu J (2020) Adaptive fuzzy backstepping control for a two continuous stirred tank reactors process based on dynamic surface control approach. Appl Math Comput 377:125138
16. Ye N, Zhao B, Salvendy G (1993) Neural-networks-aided fault diagnosis in supervisory control of advanced manufacturing systems. Int J Adv Manuf Technol 10(8):200–209

Improving the Customer Baseline Technique Based on a Learning Machine Applied to a Power System

J. Campos-Romero, J. Hernández-Núñez, and N. González-Cabrera

Abstract Demand for electricity has annualised growth due to various emerging technologies and the need to be connected to electricity to meet this consumption. In this context, it is necessary to have predictive methods of electricity consumption, such as the Customer's Baseline, which calculates the consumption pattern of users. For this reason, it is necessary to manage statistical techniques that obtain patterns as frequently as possible and reduce the estimation error. In this work we use the Naive Bayes Classifier technique based on the monitored machine learning method to improve customer baseline prediction. To validate the efficiency of the Customer Base Line—Naive Bayes Classifier model is compared to a customer Base Line—Auto-Regressive Integrated Mobile Average Model, as well as with real measures. In addition, the predictor proposed in this work is validated by indices that justify the efficiency thereof, being these the Mean Square Error, Root Mean Square Error and Mean Absolute Error. The Customer Base Line—Naive Bayes Classifier model presents an error of 0.0006%, while the Customer Base Line - Auto-Regressive Integrated Mobile Average Model presents an error of 0.014% thus justifying the superiority of the model proposed in this document. This approach was applied to a six-node system to validate its effectiveness. The results obtained from the Customer Base Line—Naive Bayes Classifier allow accurate prediction of nodal energy consumption and avoidance of unnecessary electricity consumption.

Keywords Customer base line · Naive Bayes classification · Optimal power flows · Demand response · Predictive method.

J. Campos-Romero (✉) · J. Hernández-Núñez · N. González-Cabrera
División de Ingeniería Eléctrica, UNAM, Coyoacán 04510, Mexico
e-mail: jiri_zanamicr@comunidad.unam.mx

J. Hernández-Núñez
e-mail: luis_hernandez53@comunidad.unam.mx

N. González-Cabrera
e-mail: nestorgc@fi-b.unam.mx

K. L. Flores Rodríguez et al. (eds.), *Recent Trends in Sustainable Engineering*, Lecture Notes in Networks and Systems 297,
https://doi.org/10.1007/978-3-030-82064-0_4

1 Introduction

Traditional demand-side management has been overtaken by the evolution and improvement of management in the acquisition of consumption patterns, which represents a radical change in people's quality of life, offering them an opportunity for easy access to information and control over their electricity consumption.

The demand for electricity is constantly increasing every year due to continuous improvements in the quality of life and extreme hot and cold weather conditions [1]. Requiring an accurate estimation of electricity consumption per user at distribution and power level. Demand response (DR) programs [2] is considered a promising model for improving the reliability of power systems, because it incentives consumers to change or reduce their electricity consumption during certain periods in response to technical-economic signals issued by the electricity service provider [3]. On the other hand, studies have shown that the exclusion of energy demand for price variations in the electricity market is unacceptable from an economic point of view, as it has **plagued the market with many inefficiencies** [4].

One of the DR subs-programs is the Customer Base Line (CBL) which consists on calculating the customer's baseline load consumption, and can be particularly challenging for DR users, because the daily electricity load can vary randomly and **significantly** [5]. Accurate estimation of CBL is critical to the success of DR programs because it involves the interests of multiple parties, including utilities and end-users is revised in [6, 7]. Furthermore, the use of price-based DR is supported by the extensive deployment of DR tests and the deployment of smart meters that allow **the quantification of consumer responsiveness to price signals through benchmark** estimation as can be evaluated in [8]. In [9] authors present a CBL model for measure and assessing the performance criteria of load management programs, although the solution is presented for smart grids. Time series of electricity demand patterns are the key source of information on customer consumption behaviour as is presented in [10], such information can be used in demand response programs for the prediction of user consumption, which is relevant for both the planning and operation of the energy system [11].

Demand and power generation balance using the latest load management technologies is considered an immediate requirement for demand response programs, as well as for improving the performance of electricity distribution networks [12], therefore many attempts have been made around the world to design effective DR programs for the commercial and residential industry for real customers [13]. In this context, programs that use statistical Machine Learning methods can perform better data management, which represents higher accuracy in consumption estimation and modelling [14].

The main techniques for estimating the electricity consumption in the domestic and industrial sectors are based on the least squares method [15], for example Autoregressive Integrated Moving Average (ARIMA) methodically.

Obtaining an accurate estimate of users' electricity consumption patterns will allow for reductions in energy consumption from the end-user's point of view, as well as operational certainty and a reduction in generation for the grid operator. It will even allow the user to participate in energy efficiency management programs [16].

In combination with the above, the Optimal Power Flows (OPF) are a tool to evaluate the contributions of the users in the DR programs, facilitating the addition of user-defined variables, costs and constraints to the standard OPF problem. The main objective of OPF is to minimise the cost of generation for a given load demand [17]. Finally, in accordance with the need for an accurate estimation of the CBL, this paper proposes the estimation of the Customer Base Line user demand consumption at nodal level in a more accurate way through the Naive Bayes Classifier (NBC), in order to obtain optimal responses for a 24-h horizon, in addition, this same estimation will provide signals of user consumption to the area of the system operator to plan an optimal scenario in decision making for power generation. To analyze the effectiveness of the CBL-NBC model, two scenarios are considered: first, it is compared against the ARIMA model used in [16] and second, the comparison with a weekday. Also, the OPF is considered to appreciate the nodal prices and generation as signals using the 6-node RBTS power system as a reference.

2 Methodology for the CBL and Demand Response

2.1 CBL Estimation Based on Machine Learning

In the adaptation of the NBC technique, we use hourly recorded data from weekday load consumption in total for estimating demand pattern behaviour of the next day.

The NBC is a method which assumes that the observations have some multivariate distribution given the class membership, but the predictor or entities composing the observation are independent. According to [14] in Matlab, the algorithm exploits Bayes' theorem and assumes that the predictors are conditionally independent, given the classes. Observations are assigned to the most likely class (maximum a posteriori). Explicitly, the algorithm:

1. Estimate the predictor densities within each class. We use the 'kernel' function to estimate the kernel smoothing density, specifying the Gaussian option as the kernel type, based on the following formula.

$$f(x) = \frac{1}{\sqrt{2\pi}} e^{-0.5x^2} \tag{1}$$

where x is the data predictor.

2. Models subsequent probabilities according to Bayes' rule.

$$\hat{P}(Y = k|X_1, \ldots, X_p) = \frac{\pi(Y = k)\prod_{j=1}^{p} P(X_j|Y = k)}{\sum_{k=1}^{K} \pi(Y = k)\prod_{j=1}^{p} P(X_j|Y = k)} \tag{2}$$

where Y: the random variable corresponding to the class index of an observation, $X_1, \ldots, X_P$: the random predictors of an observation, $\pi(Y = k)$: the previous probability of a class index being k.

3. Classifies an observation by estimating the posterior probability for each class and then assigns the observation of the class that produces the maximum posterior probability.

Subsequently, several indices are used to verify the efficiency of the NBC model. For example, Root Mean Square Error (RMSE), which is a measure frequently used for differences between predicted and observed values, MSE is an estimator that measures the average of errors squared, that is, the difference between the estimator and what is estimated, and the Mean Absolute Error (MAE), which is a measure of the difference between two continuous variables.

2.2 Optimal Power Flow Applied to NBC

The OPF model describes the cost minimisation of all generators connected to the power system, for all time periods t to be analysed [16–18]. The mathematical model is expressed as:

$$Min = \sum_{i=1}^{Ng} C_{i,t}(Pg_{i,t}) \tag{3}$$

subject to

$$\sum_{i=1}^{n}(Pg_{i,t} - Pd_{i,t}) = \sum_{i=1}^{n} P_{i,t}(V, \theta) \tag{4}$$

$$\sum_{i=1}^{n}(Qg_{i,t} - Qd_{i,t}) = \sum_{i=1}^{n} Q_{i,t}(V, \theta) \tag{5}$$

$$P_{ig,t}^{min} \leq P_{ig,t} \leq P_{ig,t}^{max} \tag{6}$$

$$Q_{ig,t}^{min} \leq Q_{ig,t} \leq Q_{ig,t}^{max} \tag{7}$$

$$V_{i,t}^{min} \leq V_{i,t} \leq V_{i,t}^{max} \tag{8}$$

$$P_{ig,t} \leq P_{ig,t}^{max} \tag{9}$$

where $C_{i,t}^{j}$ is the i-th generator offer of energy in the bus i, in t period, $Pg_{i,t}^{j}$ is the power generation of the bus i in the t period, Ng is the total number of generators in the system, Nb is the total buses in the system, t is the time period to be analyzed, $Pd_{i,t}^{j}$ is the active power demand in the bus i for period t,where each node i has a different consumption pattern associated with the time period t, $Qg_{i,t}^{j}$ is the reactive power generated in bus i and period t, $Qd_{i,t}^{j}$ is the reactive power demand in bus i and period t, θ is the nodal angle, $V_{i,t}^{j}$ is the voltage magnitude in node i and period t, $V_{i,t}^{j}, min$, $V_{i,t}^{j,max}$ are the minimum and maximum nodal voltage limits, $P_{im,j}^{j}$ is the line power flow through nodes i-m, n is the independent total number of elements in the entire power system (such as generators, transformers, transmission lines). Where each period t refers to the total CBL estimation behaviour load.

Equation (3) is the objective function, the constraints (4) and (5) represent active and reactive power nodal balance system, respectively, Eqs. (6) and (7) are the physical limits for the maximum and minimum active and reactive power generation respectively. Constraint (8) is the voltage magnitude limit the bus i. Equation (9) is the limit power flow.

2.3 Methodology Proposed

In this section we present the methodology proposed to compute the CBL.
Step 1. Acquisition of power consumption data from a region, obtained by means of a database implemented in a system of six RBTS nodes.
Step 2. Estimate the CBL-NBC described in the Sect. 2.1.
Step 3. Evaluate each period t obtained from the CBL-NBC in the OPF model described in Sect. 2.2.
Step 4. Analysis of the results obtained by graphical comparison of the data obtained against the actual data in horizons of 24 h, also making use of indicators that validate the efficiency of the proposed model.
Step 5. All the nodes and periods was evaluated, the process end. Other case returns Step 3.

3 Numerical Example

This section analyses the functioning of the CBL-NBC for each of the nodes of the Six bus RBTS system [18] illustrated in Fig. 1. Within this system, demand data from the electricity market regions from the Pennsylvania-New Jersey-Maryland-Interconnection (PJM) were adapted, for which it is proposed to determine the

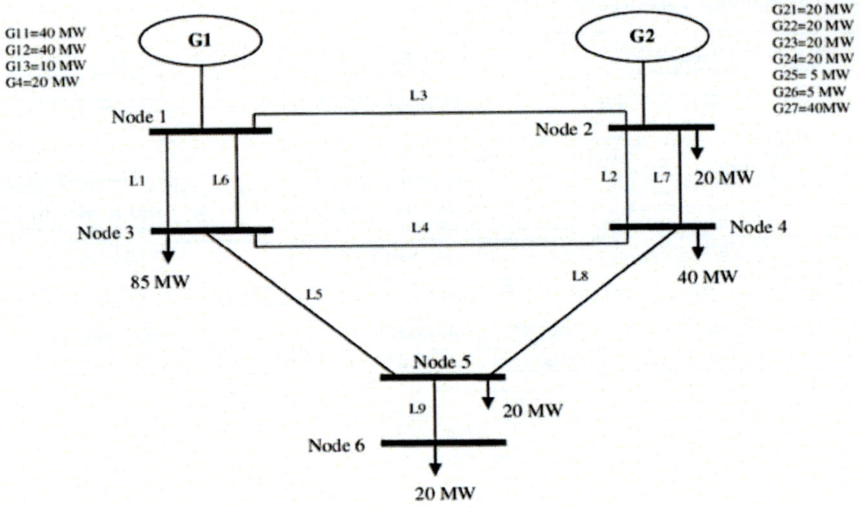

Fig. 1 Six-bus RBTS

optimal state of its operation in the electrical system, optimizing production and generation costs.

Seven days of previous consumption per zone were taken for the calculation of the time series of the CBL, so that the result obtained was compared with both the next working day and the result obtained from the ARIMA model [16] to validate the effectiveness of CBL obtained through NBC.

Figure 2 shows the behavior of the CBL-ARIMA, CBL-NBC and real data models (next day), taking the time interval from hour 10 to hour 18 as a representative section where the differences between the responses of the models can be better appreciated. You can see that the CBL-NBC and the actual data present a similar behavior, because the NBC adapts its performance to the previous data load behavior. In addition, the CBL-ARIMA model generates an estimate of energy consumption, which is higher than the real one for most of the day, so to cover this estimated consumption, it would be necessary to generate a greater amount of energy. This applies to nodes that have a load (5 of the 6 nodes), with small variations considering that the behavior of the energy is different for each of the loads.

As you can see in the figure above, with the NBC model we get a curve that almost uniformly follows the curve of the actual behavior. Linked to this and in order to ensure the superiority of the NBC model over the ARIMA model, we calculated some indicators that can be observed in Table 1.

On the other hand, the drawing of Fig. 3 of the predicted result versus Real was used to check the performance of the model, where the vertical distance of the line to any point is the error of the prediction for that point. From this same figure we can observe that most of the points are very close to the axis, so the data predicted by NBC is very well suited to the actual data.

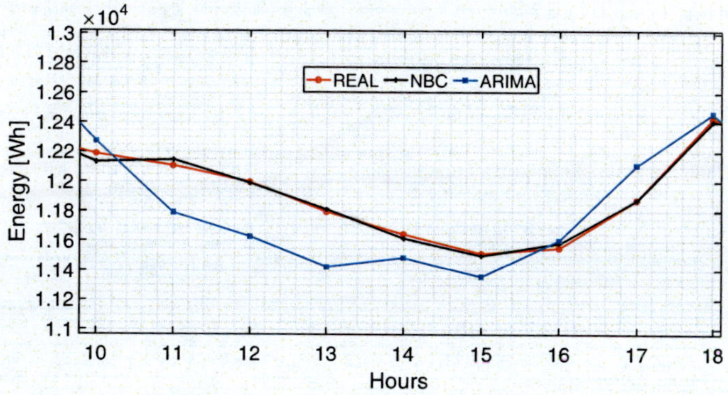

Fig. 2 CBL comparison between models for node 6 in the 10 to 18 time interval

Table 1 Indicators for model validation

	RMSE	MSE	MAE
NBC	0.064	6.8×10^{-4}	22.3118
ARIMA	1.341×10^{3}	0.014	1132.3

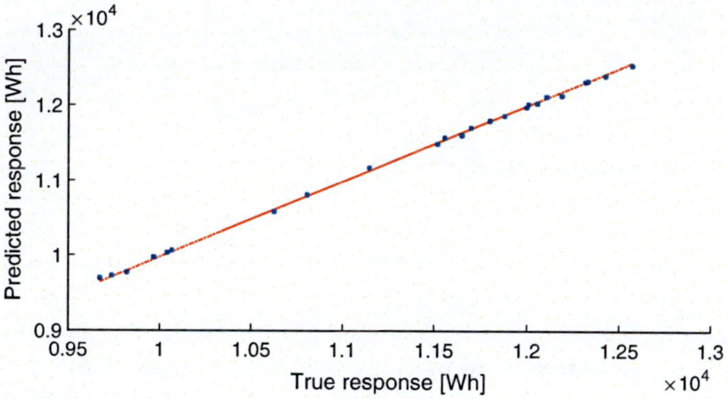

Fig. 3 NBC versus real data for node 6

Once we verified that NBC is a robust statistical method for making estimates, we made use of Optimal Power Flow Analysis (OPF) with model and real data. Given the energy price values, transmission system losses and network constraints, we obtained the nodal pricing, values for active power (P). Figure 4 presents the comparison between the hourly operating costs for CBL-NBC and CBL-ARIMA, where it can be observed that the operating cost presented by CBL-ARIMA is greater than CBL-NBC. Between hours 10 and 18 we can see how the NBC model follows almost uniformly the curve of the real data, having small variations between its values

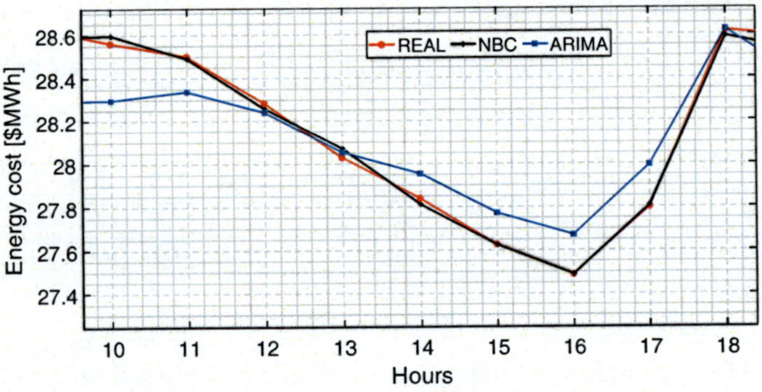

Fig. 4 Total energy cost $

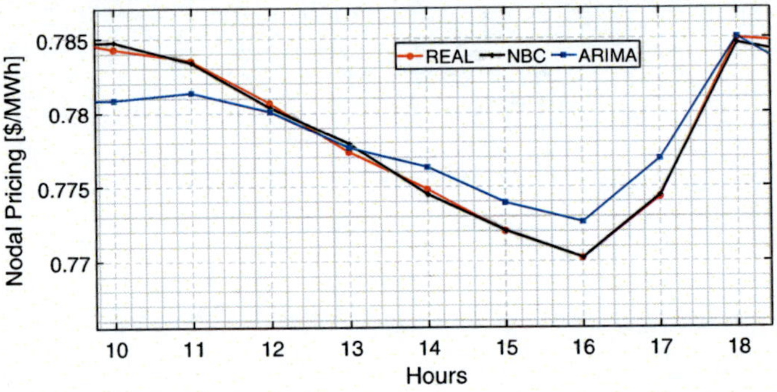

Fig. 5 Nodal pricing [$/MWh]

with a maximum difference of $0.034 in the hour 10, while the maximum difference between the actual data and ARIMA is $0.264 of the same time.

In Fig. 5, it can be noted that when comparing the prices obtained from the models, there is a small difference in tenths, which can be noticed during the time period of hour 13 to hour 18, with the most noticeable differences in hour 16 with a price of $0.0024 per MWh.

From Fig. 1, it can be seen that two of the six nodes perform the power generation function for the entire system, being nodes 1 and 2. Within the system, a total of 11 generators is contemplated, which are distributed in four generators for node 1 and 7 generators for node 2.

In Fig. 6 the active power generated in the system is represented, where it can be observed that the generation P of the ARIMA model is different compared to the estimation of the NBC model. These differences are noticeable in the following figure for the seven generators corresponding to node two, over the entire 24-h horizon.

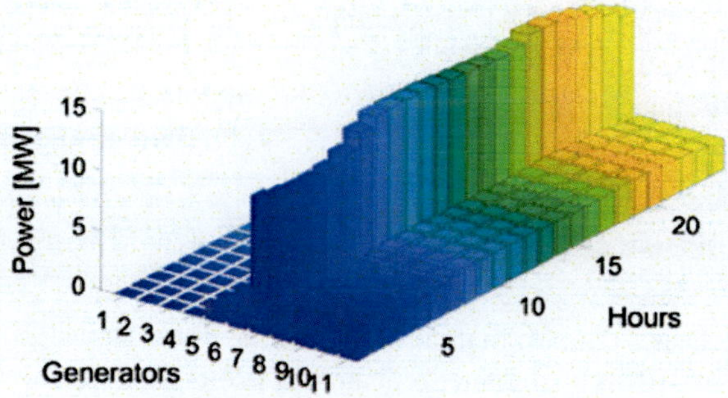

Fig. 6 Active power generation (MW)

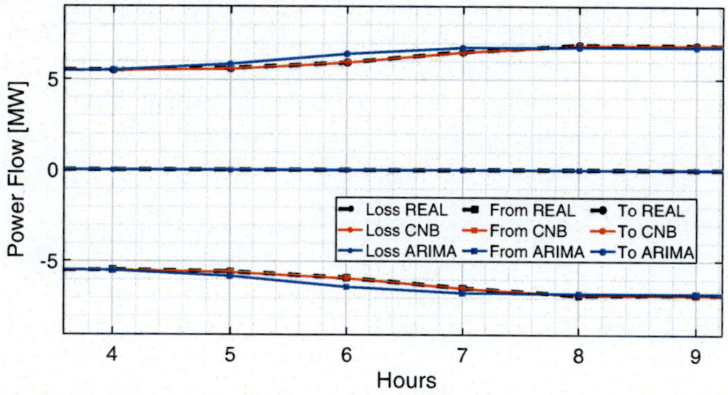

Fig. 7 Power flow for line 3

Based on the estimated power generation with both models, we note that the ARIMA model demands less generation throughout the day as mentioned above, so the demand is not met. However, by making use of the NBC model, power generation satisfies the missing energy need, that is, by having more certainty of the energy needed to supply and meet the demand, no more energy will be generated than necessary.

The system has 9 transmission lines shown in Fig. 1, these are named by the letter L and their respective line number (L1 to L9). In the lines we consider the incoming and outgoing flow P at the nodes, these help us to calculate the flow losses P of the lines.

In Fig. 7 we represent the behavior of the P flows and their losses in line 3, this is because of the 9 lines of the system, it is in line 3, during hours 4 to 9 where the disparity between the power flows predicted by the CBL-ARIMA model and the actual power flows can be observed better, being the CBL-NBC model is the best fit. In Fig. 7 we can note that the greatest difference between the methods occurs in hour 6, where the CBL-ARIMA model predicts a power output of −6.39 MW and an input of 6.42 MW while the CBL-NBC model predicts a power output of −5.91 MW and an input of 5.94 MW, comparing the predicted data with the output and power input of the actual data, which are −5.9 MW and 5.93 MW, shows that the best performance is obtained by the proposed model.

4 Conclusions

CBL-NBC as a data estimation technique has been shown to be more efficient than the CBL-ARIMA technique, reflected in principle with the values in each of the indicators, as well as the graphical validation where the actual data were compared against the predicted data, as for example in the MSE a value of 0.014% was obtained in ARIMA, while in NBC a value of 0.0006% was obtained. The estimation made by NBC makes that the generation as well as the energy flows in the lines present a response very close to that expected in the horizon of 24 h, this means that the nodal prices obtained with the NBC model maintain a small difference of approximately $0.0002 per MW with respect to the real data, while in the operating costs there is an approximate difference of $0.030, thereby demonstrating that electricity production can be covered as precisely as possible.

References

1. Lee J, Yoo S, Kim J, Song D, Jeong H (2018) Improvements to the customer baseline load (CBL) using standard energy consumption considering energy efficiency and demand response. Energy
2. Li K, Wanga F, Mi Z, Fotuhi-Firuzabadd M, Dui N, Wang T (2019) Capacity and output power estimation approach of individual behind-the-meter distributed photo-voltaic system for demand response baseline estimation. Appl Energy
3. Gagne D, Settle E, Aznar A, Bracho R (2018) Demand response compensation methodologies: case studies for Mexico. National Renewable Energy Laboratory, NREL/TP-7A40-71431 June 2018
4. Mohajeryami S, Doostan M, Schwarz P (2016) The impact of customer baseline Load (CBL) calculation methods on peak time Rebate program offered to residential customers. Department of Electrical and Computer Engineering, University of North Carolina at Charlotte, Charlotte, NC 28223. Electric Power Syst Res 137:59–65
5. Lee E, Lee K, Lee H, Kim E, Rhee W (2019) Defining virtual control group to improve customer baseline load calculation of residential demand response. Appl Energy
6. Raman G, Chih-Hsien J (2018) A hybrid customer baseline load estimator for small and medium enterprises. National University of Singapore
7. Li K, Wang B, Wang Z, Wang F, Mi Z, Zhen Z (2017) A baseline load estimation approach for residential customers based on load pattern clustering. Energy Procedia
8. Sun M, Wang Y, Teng F, Ye Y, Strbac G, Kang C, Fellow (2019) Clustering-based residential baseline estimation: a probabilistic perspective. IEEE Trans Smart Grid 10(6)
9. Sharifi R, Fathi SH, Vahidinasab V (2016) Customer baseline load models for residential sector in a smart-grid environment. Energy Rep

10. Motlagh O, Berry A, O'Neil L (2019) Clustering of residential electricity customers using load time series. Appl Energy
11. Charwand M, Gitizadeh M, Siano P, Chicco G, Moshavash Z (2020) Clustering of electrical load patterns and time periods using uncertainty-based multi-level amplitude thresholding. Electr Power Energy Syst
12. Chandran CV, Basu M, Sunderland K (2019) Demand response and consumer inconvenience 2019. In: International conference on smart energy systems and technologies (SEST), Porto, Portugal, 2019, pp 1–6. https://doi.org/10.1109/SEST.2019.8849062
13. **Wang F, Li K, Liu C, Mi Z, Shafie-Khah M, Catalão JPS (2018) Synchronous pattern match-**ing principle-based residential demand response baseline estimation: mechanism analysis and approach description. IEEE Trans Smart Grid 9(6)
14. M. Paluszek, S. Thomas, "MATLAB Machine Learning," Springer, 2017
15. Ariza Ramírez M (2013) Métodos utilizados para el pronóstico de demanda de energa eléctrica en sistemas de distribución. Universidad Tecnológica de Pereira, Colombia
16. Gonzalez Cabrera N, Gutiérrez Alcaraz G (2013) Estimation of customer base line and multi-period demand response. In: 2013 IEEE International Autumn Meeting on Power Electronics and Computing (ROPEC), Mexico City, 2013, pp 1–6. https://doi.org/10.1109/ROPEC.2013.6702709
17. **Om H, Shukla S (2015) Optimal power flow analysis of IEEE-30 bus System using soft comput-**ing techniques, vol 1, Issue-8. Department of Electrical Engineering, RCERT, Jaipur, Rajasthan, India, Nov 2015
18. Billinton R, Kumar S, Chowdhury N, Chu K, Debnath K, Goel L, Khan E, Kos P, Nourbakhsh G, Oteng-Adjei J (1989) A reliability test system for educational purposes. IEEE Trans Power Syst 4(3):1238–1244

Comparison of Euler's Backward Difference and Bilinear Transform Discretization Methods for Modeling and Simulation of a DC Motor

Filemón Arenas-Rosales, Fernando Martell-Chávez, Irma Y. Sánchez-Chávez, Rigoberto López-Padilla, and Luis Manuel Valentín-Coronado

Abstract The modelling and simulation of mechatronics systems are becoming more important with the appearance of the concept of cyberphysical systems. Virtual axes of motion can be simulated with simple DC motor models which require discretization methods. Bilinear transform method is commonly used as a precise discretization method. This study compares the Euler's backwards differences method and the bilinear transform method in both time and frequency domains for the discretization of a model of a DC motor. For the backward differences method two design criteria are used: one is the computation of the discrete pole in terms of the time constant and the sampling time, and the second is choosing a sample time corresponding to one fifteenth of the time constant. The Euler's backwards differences method is found suitable and even better than bilinear transform method for this type of application. This study shows that Euler's Backwards Discretization using a sampling time of $\tau/15$ is good to reproduce a first order dynamic response of a conventional DC motor.

Keywords Euler backward differences · Bilinear transform · DC motor modeling

F. Arenas-Rosales (✉) · F. Martell-Chávez
Centro de Investigaciones en Óptica, A.C., Unidad Aguascalientes, Aguascalientes, Mexico
e-mail: filearenas@cio.mx

F. Martell-Chávez
e-mail: fmartell@cio.mx

I. Y. Sánchez-Chávez
Universidad Politécnica de Aguascalientes, Aguascalientes, Mexico

R. López-Padilla
Universidad de Guanajuato, Guanajuato, Mexico

L. M. Valentín-Coronado
CONACYT-Centro de Investigaciones en Óptica, A.C., Unidad Aguascalientes, Aguascalientes, Mexico

K. L. Flores Rodríguez et al. (eds.), *Recent Trends in Sustainable Engineering*, Lecture Notes in Networks and Systems 297,
https://doi.org/10.1007/978-3-030-82064-0_5

1 Introduction

Mechatronic design methodologies can be used not only for the design and prototyping of physical systems but for the design of virtual machines such as virtual robots. Virtual robots and in general virtual mechatronics devices are useful for virtual model control and are becoming more used in technological trends such as digital twins and Cyber-Physical Systems (CPS) [1, 2]. 3D models of mechatronic systems are becoming more extendedly used for educational purposes such as remote training in virtual laboratories [3–5]. To design and implement virtual mechatronics devices it is necessary to develop and integrate models of the mechanisms, sensors and actuators and design a control system that can compensate the dynamical response of a real mechatronic system. Although the mechatronics systems deal with complexity the engineering research sometimes uses practical approaches to simplify the behavior of the systems, for this reason the dynamical response of mechatronic devices is often designed to have smooth and low order lineal dynamical response [6]. Mathematical models based on ordinary differential equations (ODE) are implemented to produce the signals of the mechatronic systems or components. To simulate mechatronic devices in computers it is necessary to convert the continuous time models into discrete time models, therefore discretization techniques are required. Even though the simplest differential equations can be solved exactly, in most cases a numerical treatment is necessary and the equations must be discretized to convert them into a finite system of equations that can be solved with computers [7].

One simple method for discretization is the method of finite backward differences, which replaces the differential equations in continuous time by equations of differences in discrete time. Another way to make a digital computer approach the real-time solution of differential equations is by using the Euler method, solving problems of initial numerical values (IVP) and for ordinary differential equations (ODE) [8]. Another commonly used discretization method is the bilinear transform (BLT), also known Tustin method. The BLT reduces the excess of overload during a prolonged sampling period in the digital control system [9]. The BLT is one of the most used methods due to its precise mapping of the complete left side of the s-plane into the unit circle in the z-plane and is often preferred over other discretization method such as the Euler Backward Differences (EBD) [10]. The direct current (DC) motor is a typical actuator that can be used to simulate a motion control application. DC motor has an over-damped second-order dynamic response that can be reduced to a first-order dynamic by considering only the mechanical time constant, which makes it ideal for controlling speed and position of one axis of movement, it is often used for teaching robotic control such as in [11]. Motors with linear response are useful for the implementation of virtual mechanisms, and therefore for the development of virtual robots. The modeling of DC motors is widely studied, however their numerical simulation for this type of application is not widely reported.

In this work a discretized DC motor model is proposed as a virtual actuator for the motion axes of an articulated virtual robot or other mechanisms. The analytical and numerical comparison of EBD and BLT discretization methods for a DC motor model is studied and it led to interesting results, contrary to common belief the simpler EBD method performs better under certain conditions that BLT. And additional engineering research issue addressed in this study is the selection of a suitable sampling time. The digitization of a DC motor is formulated with the EBD method, the response in both time and frequency domains is analyzed and compared with the discretization obtained with the BLT. A criterion is proposed to select the sampling frequency.

This document is structured as follows: Sect. 2 describes a method for modeling a DC motor that is used for parameter identification. The DC motor transfer function is simplified and discretized, using the BLT and EBD methods. Section 3 compares the EBD and BLT discretization methods. Section 4 presents analysis in the frequency domain of both BLT and EBD discretization. Section 5 discusses the results obtained. Conclusions are presented in Sect. 6.

2 Discretization of a DC Motor

2.1 *Mathematical Model of a DC Motor in Continuous Time and Frequency Domain*

DC motors are well known devices that can be used as actuators of many mechanisms where the speed and position of axis of movement needs to be controlled. The mathematical model of the DC motor requires two equations: the first one to model the dynamic response of the electric circuit, and the second one to model the dynamic response of the mechanical device. These equations are based on the Kirchhoff's circuit laws and Newton's motion laws [12]. Figure 1 shows the electromechanical model of a DC motor.

The model illustrated in Fig. 1 has the following electrical and mechanical parameters; V = voltage, ω = angular speed, $i(t)$ = armature current, L = armature inductance, R = armature resistance, J = rotor inertia, V_b = back electromagnetic field (EMF), B = friction coefficient, τ_M = DC motor load torque.

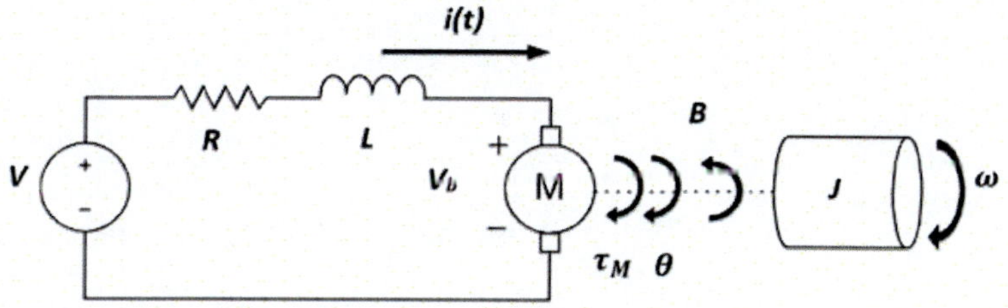

Fig. 1 Circuit of the DC motor

Applying Kirchhoff's laws, a first order differential equation, Eq. (1), can be express for the electric circuit;

$$V(t) = R \cdot i(t) + L\frac{di(t)}{dt} + K_b \cdot \omega(t) \tag{1}$$

the back emf is;

$$V_b(t) = K_b \cdot \omega(t) \tag{2}$$

Equation (3) is obtained substituting Eq. (2) into Eq. (1);

$$V(t) = R \cdot i(t) + L\frac{di(t)}{dt} + V_b(t) \tag{3}$$

For the mechanical part of the DC motor, another first order differential equation is considered, Eq. (4), where $\tau_L =$ Disturbance variable;

$$\tau_M - \tau_L = J\frac{d\omega(t)}{dt} + B \cdot \omega(t) \tag{4}$$

Considerate $\tau_L = 0$.
where;

$$\tau_M(t) = K_t \cdot i(t) \tag{5}$$

Substituting Eq. (5) in Eq. (4);

$$K_t \cdot i(t) = J\frac{d\omega(t)}{dt} + B \cdot \omega(t) \tag{6}$$

Equations (3) and (6) can also be represented in the frequency domain by applying the Laplace transform:

$$V(s) = R \cdot i(s) + Ls \cdot i(s) + V_b(s) \tag{7}$$

$$K_t \cdot i(s) = Js \cdot \omega(s) + B \cdot \omega(s) \tag{8}$$

where;

$$i(s) = \frac{V(s) - V_b(s)}{R + Ls} \tag{9}$$

Substituting Eq. (9) in Eq. (8):

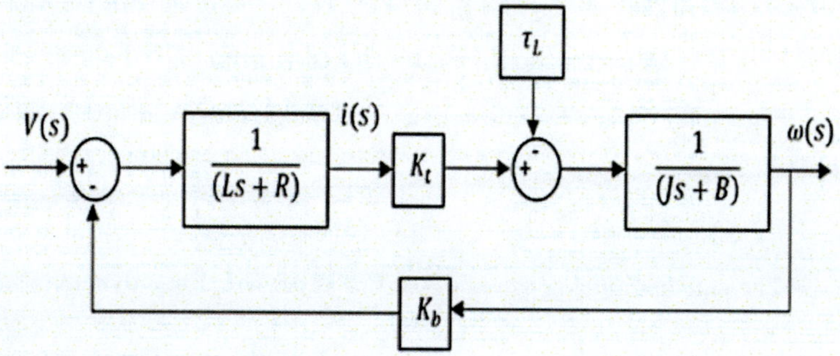

Fig. 2 Block diagram of a DC motor

$$K_t \cdot \frac{V(s) - V_b(s)}{R + Ls} = Js \cdot \omega(s) + B \cdot \omega(s) \tag{10}$$

Figure 2 shows a typical block diagram of the mathematical model of a DC motor, representing in the left side the electromagnetic process of the stator, and in the right side the mechanical process of the rotor, where the electric circuit current is converted to torque, the load torque is subtracted, and the remainder torque is an input to the mechanical circuit, having a feedback of the induced voltage [13].

From the block diagram of Fig. 2, where K_t = torque constant, K_b = EMF constant, the transfer function of interest is:

$$G(s) = \frac{\omega(s)}{V(s)} = \frac{K_t}{JLs^2 + JRs + LBs + BR + K_t \cdot K_b} \tag{11}$$

The second order system can be reduced to a first order system considering the fact that the mechanical time constant has an order of difference with respect to the electrical time constant. Showing the electrical circuit using only a constant gain (K_e), a simplified block of the DC motor can be obtained corresponding to a first order system:

From Fig. 3, the input–output relation is given by:

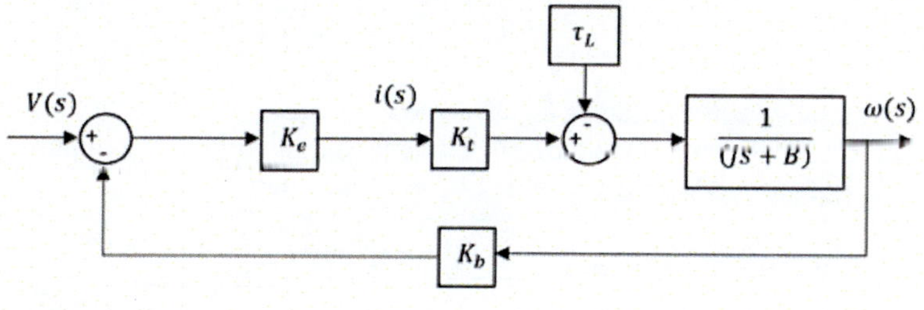

Fig. 3 Simplified block diagram of DC motor

$$G(s) = \frac{\omega(s)}{V(s)} = \frac{K_e \cdot K_t}{Js + B + K_e \cdot K_t \cdot K_b} \tag{12}$$

Equation (12) can be adjusted to match a first order transfer function:

$$G(s) = \frac{\omega(s)}{V(s)} = \frac{K}{\tau s + 1} \tag{13}$$

The gain and time constant of the first order simplified dynamics are:

$$K = \frac{K_e \cdot K_t}{B + K_e \cdot K_t \cdot K_b} \tag{14}$$

$$\tau = \frac{J}{B + K_e \cdot K_t \cdot K_b} \tag{15}$$

2.2 Model Discretization by Bilinear Transform (BLT)

For computational simulation difference equations are needed, Bilinear Transform is a discretization method conventionally used to convert a transfer function in Laplace domain to z-domain transfer function from which difference equations can be obtained.

The BLT it is used in the design of digital filters to obtain the desired response characteristics [14]. The BLT performs a mapping between the s-plane and z-plane according to Eq. (16) where T is the sampling period of the discrete time control system under consideration:

$$s = \frac{2}{T}\frac{(z-1)}{(z+1)} \tag{16}$$

Applying bilinear transform to Eq. (13):

$$G(z) = \frac{K}{\frac{2\cdot\tau}{T}\left(\frac{z-1}{z+1}\right) + 1} \tag{17}$$

Input–output variables are changed to "u" and "y" to match the nomenclature of discrete differences with are conventional terminology for an input–output system.

Equation (18) in the domain of z-plane is obtained;

$$\frac{\omega(s)}{V(s)} = \frac{y(s)}{u(s)} = \frac{K\left(1 + s^{-1}\right)}{(1+\sigma) + s^{-1}(1-\sigma)} \tag{18}$$

where:

$$\sigma = \frac{2\tau}{T} \tag{19}$$

By applying the inverse z transform an expression for the output is obtained;

$$y_k = \frac{(\sigma - 1)y_{k-1} + K(u_k + u_{k-1})}{1 + \sigma} \tag{20}$$

2.3 *Discretization by Euler's Backward Differences Method*

The previous discretization requires to obtain transfer functions in s and z domains. An alternative way to obtain the difference equations is to apply directly in the time domain the numerical differentiation defined by EBD method.

Expressions of First Differences backward;

$$f'(x) = \frac{f(x_n) - f(x_{n-1})}{\Delta T} \tag{21}$$

The first order ODE for the DC motor considering $y = \omega$ and $V = u$ for consistency with the nomenclature used in a discretization:

$$\tau \frac{dy(t)}{dt} + y(t) = K \cdot u(t) \tag{22}$$

Applying EBD to the Eq. (22):

$$\tau \cdot \left(\frac{y_n - y_{n-1}}{T}\right) + y_n = K \cdot u_n \tag{23}$$

Equation (23) can be written as:

$$y_n - y_{n-1} + y_n \cdot \left(\frac{T}{\tau}\right) = K \cdot u_n \cdot \left(\frac{T}{\tau}\right) \tag{24}$$

Rearranging Eq. (24):

$$y_n = y_{n-1} \cdot \left(\frac{\tau}{\tau + T}\right) + K \cdot u_n \cdot \left(\frac{T}{\tau + T}\right) \tag{25}$$

The following relations are defined to be parametrized. The difference equation obtained directly with the EBD method can have the parameters α, τ and T in discrete time, defining the following relations:

$$\alpha = \frac{\tau}{\tau + T} \tag{26}$$

$$(1 - \alpha) = \frac{T}{T + \tau} \tag{27}$$

The discrete model can be expressed as:

$$y(n) = \alpha \cdot y_{k-1} + (1 - \alpha) \cdot K \cdot u_n \tag{28}$$

3 Comparison of EBD and BLT Discretization Methods

This study compares two discretization methods for the model of a DC motor suitable to be used in the implementation of virtual mechanisms. The evaluation of the previously obtained difference equations Eqs. (20) and (25) is performed to compare the two discretization methods under study. A sampling time of $\tau/15$ is selected and this selection is explained later in this section.

3.1 Comparison in the Time Domain

To compare both the discretization obtained with EBD and the discretization obtained with BLT, a unit step input is considered. The continuous time step response is:

$$y(t) = 1 - e^{\frac{-t}{\tau}} \tag{29}$$

The exact solution for the discrete time step response is;

$$y(n) = 1 - e^{\frac{-nt}{\tau}} \tag{30}$$

The simulations for the discrete time systems from EBD and BLT are carried out using the sampling period T = $\tau/15$. The response values are shown in Table 1. The columns refer to the exact solution, the bilinear transform (BLT) and the backward differences (EBD) approximation; and the errors between BLT and the exact solution, and between EBD and the exact solution. The error of the analyzed methods and of the integral of the error are shown in Figs. 4 and 5, respectively.

Table 1 shows the values obtained using a sampling time T = $\tau/15$, for $n = 1$ to 20. As the number of iterations increases, approximations get closer to the exact solution. The error generated by the BLT and EBD methods is also compared and analyzed with respect to the exact solution, showing that the EBD method has a lower error compared to the BLT method. A more detailed observation can be seen if the

Table 1 Numerical values and errors for sampling period $T = \tau/15$

n	Exact solution	Bilinear transform		Backward differences	
	y_n	$y(k_n)$	e_r	$y(k_n)$	e_r
1	0.0091	0.0045	0.0045	0.0088	0.0002
2	0.0176	0.0134	0.0042	0.0171	0.0005
3	0.0256	0.0216	0.0039	0.0249	0.0007
4	0.0331	0.0294	0.0037	0.0322	0.0009
5	0.0401	0.0366	0.0034	0.0390	0.0010
6	0.0466	0.0434	0.0032	0.0454	0.0012
7	0.0528	0.0497	0.0030	0.0514	0.0013
8	0.0585	0.0557	0.0028	0.0571	0.0014
9	0.0639	0.0612	0.0026	0.0624	0.0015
10	0.0689	0.0664	0.0024	0.0673	0.0015
11	0.0736	0.0712	0.0023	0.0719	0.0016
12	0.0779	0.0758	0.0021	0.0763	0.0016
13	0.0821	0.0800	0.0020	0.0804	0.0016
14	0.0859	0.0840	0.0019	0.0842	0.0016
15	0.0895	0.0877	0.0017	0.0878	0.0016
16	0.0928	0.0912	0.0016	0.0912	0.0016
17	0.0960	0.0944	0.0015	0.0943	0.0016
18	0.0989	0.0975	0.0014	0.0973	0.0016
19	0.1017	0.1003	0.0013	0.1000	0.0016
20	0.1043	0.1030	0.0012	0.1026	0.0016

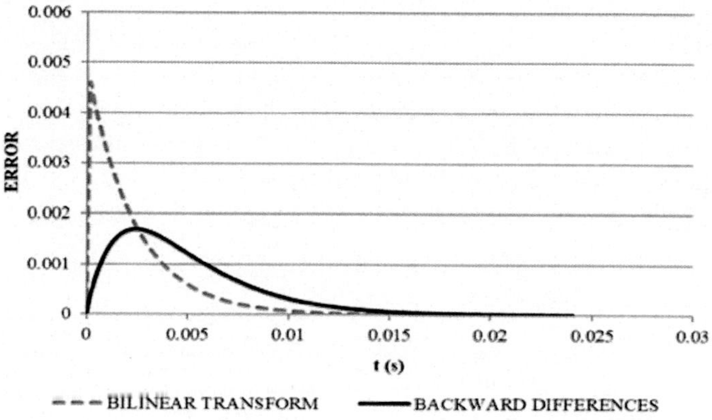

Fig. 4 Error with $T = \tau/15$

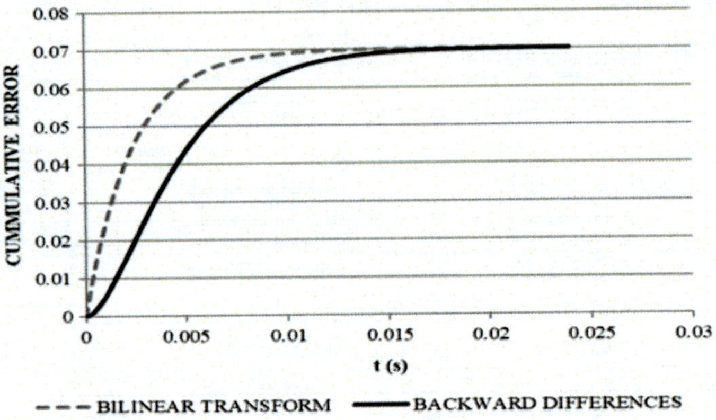

Fig. 5 Integral of the error with $\mathrm{T} = \tau/15$

error values are charted. Figure 4 shows that for time values $t < \tau$, the EBD method has a smaller error compared to the BLT method, but for time values $t > \tau$ the BLT method gets the lower error.

Figure 5 shows values of the integration of the error (cumulative error). It can be seen that in the transient response the BLT accumulates a greater value, than the EBD but in long term cumulative errors tend to converge.

Another interesting analysis can be done selecting different sampling periods. Table 2 shows the obtained values for the numerical evaluation at the sample $n = \tau/T$ for the exact solution, the EBD and BLT methods, and for each one of the following sample periods: $\mathrm{T} = \tau/5$, $\mathrm{T} = \tau/10$, $\mathrm{T} = \tau/15$, $\mathrm{T} = \tau/20$ and $\mathrm{T} = \tau/25$. The response values and errors of EBD and BLT are charted respectively in Figs. 6 and 7.

In Fig. 6 it can be seen that for higher sampling periods ($\mathrm{T} = \tau/5$, $\mathrm{T} = \tau/10$) EBD method has a smaller difference with respect to the exact solution than BLT method. For lower sampling periods ($\mathrm{T} = \tau/15$, $\mathrm{T} = \tau/20$ and $\mathrm{T} = \tau/25$) the amplitude values of EBD and BLT gets very close and with the same deviation with the exact solution, a detailed observation is further needed.

Table 2 Numerical values and errors for sampling periods $\mathrm{T} = \tau/5$, $\mathrm{T} = \tau/10$, $\mathrm{T} = \tau/15$, $\mathrm{T} = \tau/20$, $\mathrm{T} = \tau/25$ and evaluated in $n = \tau/T$

$\frac{\tau}{n}$	Exact solution	Bilinear transform		Backward differences	
	y_n	$y(k_n)$	e_r	$y(k_n)$	e_r
$\tau/5$	0.0895	0.0839	0.0056	0.0847	0.0048
$\tau/10$	0.0895	0.0868	0.0027	0.0870	0.0025
$\tau/15$	0.0895	0.0877	0.0017	0.0878	0.0016
$\tau/20$	0.0895	0.0882	0.0013	0.0882	0.0012
$\tau/25$	0.0895	0.0884	0.0010	0.0885	0.0010

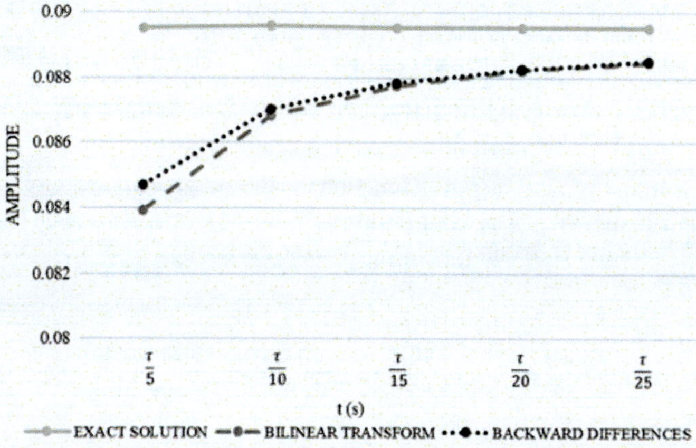

Fig. 6 Response values for the different sampling periods and evaluated in n = τ/T

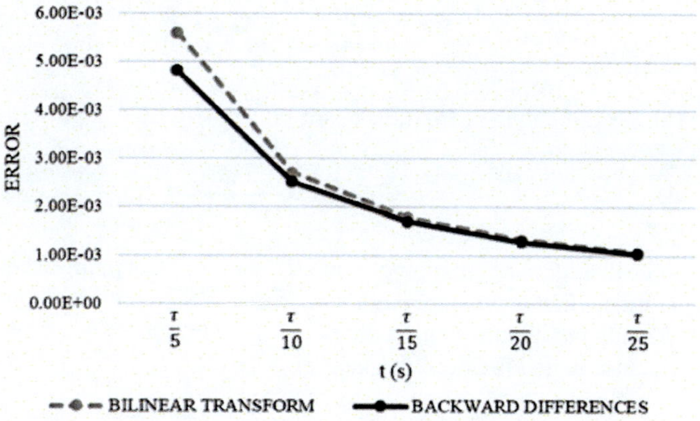

Fig. 7 Error values for the different sampling periods and evaluated at n = τ/T

In Fig. 7 a comparisson of the errors of EBD and BLT with respect to the exact solution is shown. For a sampling time of $\tau/5$ the error of BLT shows a 6.25% compared to EBD that is only 5.38%, this indicates that EBD is more precise than BLT. Considering now a sampling time of $\tau/15$, both the EBD and BLT methods have an error slightly less than 1.89%. As the sampling time is reduced, the error values get closer and converge to the same value.

4 Comparison in the Frequency Domain

4.1 BLT Analysis in the Frequency Domain

The w-transform is a mapping function for the z-domain. The w-plane is used instead of the s-plane, the w-transform is used for the analysis of discrete time systems. This transformation is detailed in [15]. The w-transformation is a bilinear transformation given by;

$$z = \frac{\left(1 + \frac{T}{2} \cdot w\right)}{\left(1 - \frac{T}{2} w\right)} \tag{31}$$

Substituting Eq. (31) in Eq. (18) to obtain;

$$G(w) = \frac{K\left(\frac{1+\frac{T}{2}\cdot w}{1-\frac{T}{2}\cdot w} + 1\right)}{\left(\frac{1+\frac{T}{2}\cdot w}{1-\frac{T}{2}\cdot w}\right) \cdot (1+\sigma) + (1-\sigma)} \tag{32}$$

The simplification of Eq. (32) result in:

$$G(w) = \frac{K}{\tau \cdot w + 1} \tag{33}$$

Since the frequency mapping from s to z and the w mapping are defined by bilinear transforms, a first order system is obtained in w. Therefore, the frequency response for the original model in s-plane is the same as the frequency response in the frequency w (w is also called fictitious frequency).

4.2 EBD Analysis in the Frequency Domain

The analysis in the w-plane is done for discrete transfer functions in z, to obtain a transfer function in z for the first order transfer function in s a transformation is required, the relation of s and z for the backward differences discretization is defined by Eq. (34) or Eq. (35);

$$s = \frac{z - 1}{zT} \tag{34}$$

$$s = \frac{1 - z^{-1}}{T} \tag{35}$$

Substituting Eq. (35) into Eq. (13):

$$G(z) = \frac{K}{\tau \cdot \left(\frac{1-z^{-1}}{T}\right) + 1} = \frac{K \cdot z}{z \cdot \left(1 + \frac{\tau}{T}\right) - \frac{\tau}{T}} \tag{36}$$

Now, the w transformation Eq. (31) can be applied to Eq. (36):

$$G(z) = \frac{K \cdot \left(\frac{1+\frac{T}{2}\cdot w}{1-\frac{T}{2}\cdot w}\right)}{\left(\frac{(1+\frac{T}{2}\cdot w)\cdot(1+\frac{\tau}{T})}{1-\frac{T}{2}\cdot w}\right) - \frac{\tau}{T}} \tag{37}$$

Simplifying the Eq. (37) to obtain;

$$G(w) = \frac{K \cdot (Tw + 2)}{w \cdot (2\tau + T) + 2} \tag{38}$$

Figure 8 shows the Bode Diagrams with the frequency response of the transfer functions of Eqs. (33) and (38), corresponding to the frequency responses of BLT and EBD respectively. A comparison of the frequency response is done and it shows similar responses of both magnitude and phase in the bandwidth of interest, this is below the cut-off frequency of the system under analysis. A different magnitude and phase is noticed but for frequencies higher than the cut-off frequency, this is expected because of the existence of the zero. This frequency analysis validates the equivalence of both discretization methods.

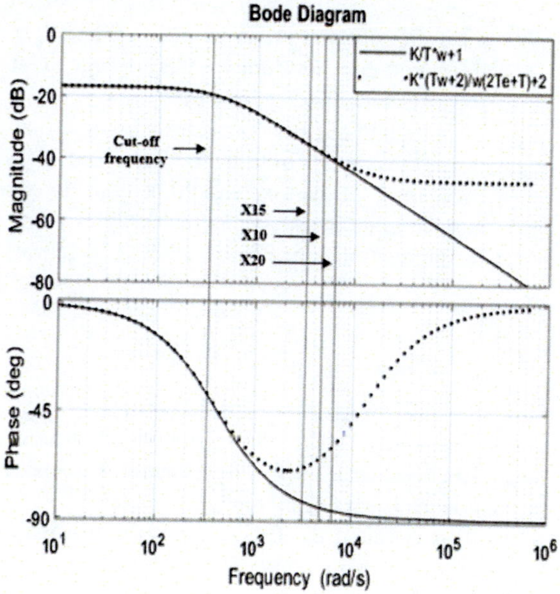

Fig. 8 Comparison of BLT and EBD discretization both in magnitude and phase

The cut-off frequency of the model under study is 327 rad/s, considering a sampling frequency given by $\tau/10$ a bandwidth of 3270 rad/s is obtained, below this sampling frequency the magnitude and phase of both systems BLT and EBD coincides. Now considering a sampling of $\tau/15$ generates a bandwidth of 4905 rad/s, it can be observed that the magnitude between both systems is still fairly close but now they have a 45° ($\pi/4$) differences in the phase. Finally, considering a sampling frequency of $\tau/20$ that corresponds to a bandwidth of 6540 rad/s, the magnitude and phase begins to differentiate for higher frequencies. From this frequency analysis it can be said that the EBD discretization have better both gain margin and phase margin than BLT discretization.

5 Discussion of Results

The EBD method gets results closer to the exact solution. The EBD method is simpler since the conversion can be done directly from the ODE by just considering the relationship $\alpha = \frac{\tau}{\tau+T}$ for EBD.

Any sample time between $\tau/10$ and $\tau/20$ is good for both methods, in particular in the EBD method meets this criterion, it is already comparable to the BLT method having a better response in discrete time for values less than τ. As the sampling frequency increases (reduction of the sampling time), the error between EBD and BLT is reduced.

The BLT method and EBD method have fairly equivalent frequency responses between $\tau/10$ and $\tau/20$, the frequency response is different for higher frequencies when the sampling time is greater than $\tau/20$, from the analysis performed in both time and frequency domains it is possible to establish that $\tau/15$ is a good specification for a discretization sample time.

Although the BLT provides complete mapping from the left side of the s-plane inside the unitary circle in the z-plane and this is the reason this digitalization is widely used, the EBD method should be considered as a good option since the inaccuracy and damping introduced in the complex plane mapping ends up being beneficial for the simulation, since it guarantees the stability of the system.

6 Conclusion

This study shows a comparison of two popular discretization methods, such as EBD and BLT, the discretization was required for the computational simulation of a conventional DC motor.

It is found the EBD discretization for first order systems can be done by just applying two simple design criteria: the first is the discretization of the first order ODE by just using $\alpha = \frac{\tau}{\tau+T}$, second choosing a sample time corresponding to step size of $\tau/15$.

When the EBD method is compared with the BLT method in discrete time transient response, it shown a better response for time values before τ, and its error is less than the obtained by BLT. From the frequency response analysis performed in this study it is found that a sampling frequency corresponding to discrete step size of $\tau/15$ is an adequate to keep the equivalence of the EBD with regards to BLT.

From this study we can conclude that the EBD method is suitable and even better than BLT for applications where a first order dynamic response is required to be simulated like the modeling and simulation of DC motors.

References

1. Herbus K, Ociepka P (2015) Integration of the virtual 3D model of a control system with the virtual controller. IOP Conf Ser Mater Sci Eng 95:012084
2. Smerpitak K et al (2007) A technique for mechanical adjustment of DC motor by virtual model. In: International conference on control, automation and systems 2007, 17–20 Oct 2007, COEX, Seoul, Korea
3. Familia R (2005) A virtual laboratory for cooperative learning of robotics and mechatronics. In: Dominican Republic; ITHET 6th annual international conference T2B-20. 0-7803-9141-1/05/. IEEE
4. Candelas FA et al (2003) A virtual laboratory for teaching robotics. Int J Eng Ed 19(3):363±370
5. Bargsten V, Zometa P, Findeisen R (2013) Modeling parameters identification and model-based control of a lightweight robotic manipulator. In: IEEE international conference on control applications (CCA). 978-1-4799-1559-0/13/
6. Rengifo CF et al (2017) A performance comparison of nonlinear and linear control for a DC series motor. Ciencia en Desarrollo, vol. 8, no. 1. Enero-Junio de 2017, pp 41–50. ISSN 0121-7488
7. Scherer POJ (2013) Computational physics; simulation of classical and quantum systems. Springer, Germany, 255 p
8. Amirul Islam M (2015) Accuracy analysis of numerical solutions of initial value problems (IVP) for ordinary differential equations (ODE). IOSR J Math (IOSR-JM) 11(3):18–23. Ver. III (May–June 2015). e-ISSN: 2278-5728, p-ISSN: 2319-765X
9. Zhai G et al (2013) An extension of generalized bilinear transformation for digital redesign. Int J Innov Comput Inf Control. ICIC International. ISSN 1349-4198
10. De Keyser R et al (2018) An efficient algorithm for low-order direct discrete-time implementation of fractional order transfer functions. https://doi.org/10.1016/j.isatra.2018.01.026. Published by Elsevier Ltd.
11. Craig JJ (2006) Robotica Mexico. Pearson Education
12. Chaturtvedi DK (2010) Modeling and simulation of systems, using MATLAB and simulink. CRC Press, USA, pp 84–86
13. Gaviño RH (2010) Introducción a los sistemas de control; conceptos, aplicaciones y simulación con matlab. Pearson, México, 95 p
14. Jiang Y, Hu X, Wu S (2014) Transformation matrix for time discretization based on Tustin's method. Math Probl Eng 2014. Hindawi Publishing Corporation. Article ID 905791, 9 p
15. Ogata K (1996) Sistemas de Control en Tiempo Discreto. Prentice Hall Hispanoamericana, México, 228 p

Further Results on Modeling and Control of a 3-DOF Platform for Driving Simulator Using Rotatory Actuators

Iván Cañedo Farfán, Roberto Carlos Ambrosio Lázaro, and José Fermi Guerrero Castellanos

Abstract The present work aims to develop a mathematical model and position control algorithm of a 3-RRS (revolute, revolute, spherical) parallel manipulator, for driving simulators. For this purpose, the inverse and forward kinematics are considered. Furthermore, the actuator model, DC-Motors, is taken into account. A simulation-based approach is presented using MATLAB/Simulink with Simscape complement for the mechanical constraints. Then, each actuator's position control is performed using the ADRC methodology, which provides robustness and simplicity in its implementation. Simulation results validate the proposal and the overall performance of the system is measured and compared with a PID-based solution using the ISE index.

Keywords Driving simulator · Parallel manipulator · Inverse kinematics · ADRC · 3-RRS

1 Introduction

Driving simulators are tools that allow performance of various tasks, like training new drivers or carrying out tests in a controlled environment that would otherwise be dangerous or expensive to perform in a real vehicle. Therefore, universities and industrial laboratories have been using driving simulators since the early 70s [1].

Nowadays, driving simulators have become more accessible, given that computational capabilities have been increasing in power and decreasing in price yearly. Thus, many configurations have been developed, ranging from static (0-DOF) sim-

I. Cañedo Farfán (✉) · R. C. Ambrosio Lázaro · J. F. Guerrero Castellanos
Benemérita Universidad Autónoma de Puebla, Puebla PUE, Mexico
e-mail: ivan.canedo@alumno.buap.mx

R. C. Ambrosio Lázaro
e-mail: roberto.ambrosio@correo.buap.mx

J. F. Guerrero Castellanos
e-mail: fermi.guerrero@correo.buap.mx

K. L. Flores Rodríguez et al. (eds.), *Recent Trends in Sustainable Engineering*, Lecture Notes in Networks and Systems 297,
https://doi.org/10.1007/978-3-030-82064-0_6

ulators to systems with 9-DOF or more. The minimum DOF needed to determine the driving simulator's validity is closely related to the type of task that the driving simulator will be used for [2]. Ergo, a compromise between fidelity of the simulation and the minimum DOF must be reached. Therefore, a modeling and simulation of the control algorithm and inverse kinematic of a 3-DOF parallel platform using ADRC is presented in this work.

1.1 Motion-Based Driving Simulators

Recently, there has been an increasing interest in research for driving simulators with different configurations for automotive applications. The most widely accepted configurations are parallel manipulators (PM), which can be categorized by the type of joint on the kinematic chains that move the platform. The most popular configuration is the SP configuration, where the kinematic chains are conformed by two spherical passive joints and an actuated prismatic joint, this configuration is the best documented so far, but it is not the only existing configuration, other popular configurations include the RRS, RSS and RRR, where R stands for revolute joint.

To control the position of a moving platform two components are needed: the model of the platform, and the control of the actuators, those components have the goal of tracking a reference. This reference is a signal that contains the desired position of the platform. Since the platform being considered is meant to be used in a driving simulator, the reference signal is time variant, not deterministic and aperiodic, with low and high frequency components. Therefore, the validation of the tracking system can be more complex than the validation of a reference tracking system with a well-known reference signal.

1.2 Previous Works

In this section, a review of previous works regarding the modeling and control of parallel manipulators is presented. There are, mainly, three subjects of study regarding parallel manipulators: (1) the inverse model, (2) the forward model, and (3) the workspace calculation. When the platform uses an inverse model, the reference is the desired platform position and the output is the actuator's desired position. In a forward model, the reference is the desired actuator position and the output is the platform position. The workspace calculation consists in the determination of all possible positions that the platform can reach considering the mechanical constrains of the specific configuration.

Regarding the 3-DOF parallel manipulators, like its 6-DOF counterpart, the most common configuration for the 3-DOF parallel manipulators is based on actuated prismatic joints, but the main difference is that instead of using an SP confirmation, the most common configuration is RPS (revolute, prismatic, spherical) [3, 4]. Thus, regarding the modeling of the platform, there are two approaches to the problem, the inverse kinematics, and the dynamics modeling, in reference [5] a comparison between the kinematics and dynamics of an RRS and an RSR configuration is performed considering the complexity, workspace, and singularities of every model; in [6] a velocity an acceleration analysis was performed for the RRS configuration, in [7] the inverse kinematics was solved using a conformal geometric algebra approach and the algorithm was validated with a small scale experimental platform.

2 Problem Statement

Medium-scale motion-based commercial driving simulators using a 3-DOF parallel manipulator with an RRS configuration and DC gearmotors as actuatora have recently become popular due to their capability of providing a motion feedback cue to the simulator operator. The advantage of a more affordable price than its 6–9 DOF counterpart, makes it more attractive for research in the field of driving simulators, nevertheless, real performance of those systems along with modeling and control strategies used are not openly available.

Driving simulators are complex systems with many interrelated components, therefore, to represent the system comprehensively it is necessary to simplify some components, thus a block diagram of a driving simulator is shown in Fig. 1, in this diagram the system is represented as a control problem focusing on the control of the platform's movement. To generate the platform reference position, it is necessary to have the acceleration and angular velocity of the simulated vehicle and process them in a motion cueing algorithm (MCA). The MCA is designed to transform the velocity and acceleration experimented by the vehicle to movements of the platform that have an equivalent stimulation in the vestibular organ of the driving simulator operator. With the desired position of the platform obtained from the MCA, the next step is controlling of the platform per se, the proposed control comprises the inverse kinematics of the platform and the control of each motor. To measure the

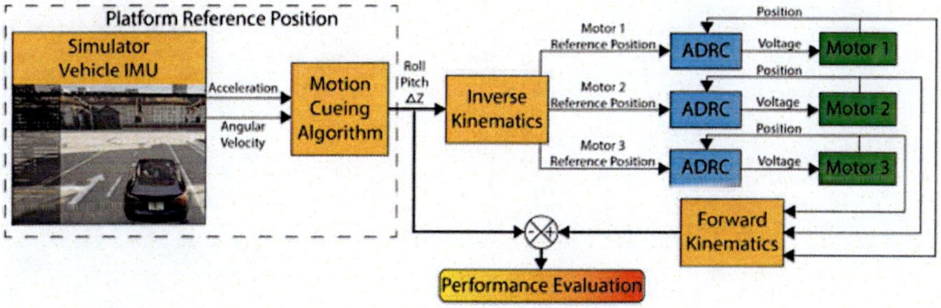

Fig. 1 Block diagram of the architecture of a driving simulator

performance of the control, the reference position is compared to the position of the platform obtained from forward kinematics.

As shown in Sect. 1.2, regarding the 3-DOF RRS parallel manipulator, most of the published works deal with the modeling of the platform. Nevertheless, the model of the motor actuating the revolute joint is rarely mentioned, Therefore, a simulation-based research can provide valuable information about the performance of a 3-DOF RRS parallel manipulator.

3 Modeling

The 3-DOF RRS parallel manipulator that is the focus of this research is shown in Fig. 2. Composed of two platforms and tree legs, the coordinate system *B-XYZ* is attached to the center of the base platform and it is considered to remain static. The coordinate system *P-xyz* is attached to the center of the moving platform, B_i (i = 1,2,3) are the connecting points to the base platform and similarly P_i are the connecting points of the moving platform, R_i are the union points for the passive revolute joints, L and l are the length of the links between $B_i - R_i$ and $R_i - P_i$ respectively, ℓ is the distance between B_i and P_i, the angle λ is the angular position of the actuated revolute joint B_i with respect to the vertical. The degrees of freedom of the system are roll, pitch and heave ($\phi, \theta, \Delta Z$).

3.1 Inverse Kinematics

Given that the separation between every point is 120°, the position for the union points of the base and platform are defined in (1) and (2).

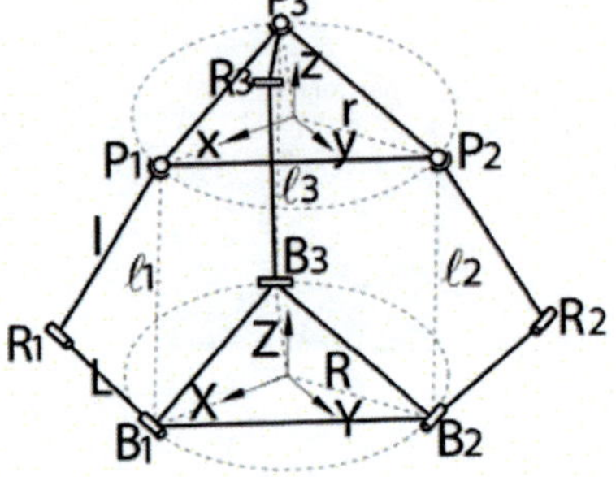

Fig. 2 The schematic diagram of the 3-RRS parallel manipulator

$$B_i = \begin{bmatrix} B_x \\ B_y \\ B_z \end{bmatrix} = \begin{bmatrix} R\cos a_i \\ R\sin a_i \\ 0 \end{bmatrix} \tag{1}$$

$$P_i = \begin{bmatrix} P_x \\ P_y \\ P_z \end{bmatrix} = \begin{bmatrix} r\cos a_i \\ r\sin a_i \\ 0 \end{bmatrix} \tag{2}$$

with:

$$a_1 = 0°, a_2 = 120°, a_3 = 240° \quad \text{and} \quad i = (1, 2, 3) \tag{3}$$

The next step is to calculate the new position of the platform, in order to do this, a translation vector T and a rotation matrix R are defined as a function of the three degrees of freedom of the platform. The translation vector is shown in (4) where the translation in the X and Y axis must be 0 and the translation on the Z axis (ΔZ) is defined as the sum of the desired longitudinal translation and the platform height H, the height is set arbitrarily choosing a desired "home" position, on the other hand the rotation matrix is shown in (5) as a function of the desired orientation of the platform, where the rotation around the X axis and Y axis (roll, pitch) are ϕ and θ, the rotation around the Z axis is not taken into consideration, given that $\psi = 0$.

$$T = \begin{bmatrix} T_x \\ T_y \\ T_z \end{bmatrix} = \begin{bmatrix} 0 \\ 0 \\ H + \Delta Z \end{bmatrix} \tag{4}$$

$$R = R_x(\phi)R_y(\theta)R_z(0) = \begin{bmatrix} \cos\theta & \sin\phi\sin\theta & \cos\phi\sin\theta \\ 0 & \cos\phi & \sin\phi \\ -\sin\theta & \cos\theta\sin\phi & \cos\phi\sin\theta \end{bmatrix} \tag{5}$$

Next, the desired position of the platform points P_d is calculated using (6) with i = (1, 2, 3)

$$P_{di} = RP_i + T - B_i \tag{6}$$

Once the desired P_d is obtained, the next step is to calculate the angle λ, thus, three vertical planes are defined such that the points of P are on those planes. A graphical representation of this concept is shown in Fig. 3.

With the new planes defined, it is possible to model the kinematic legs, as shown in Fig. 4. And by using this geometric representation we can now calculate the angle λ, which is obtained by calculating the angles α, β an the line ℓ using (7) through (10).

$$\ell_i = \| P_{di} \| \tag{7}$$

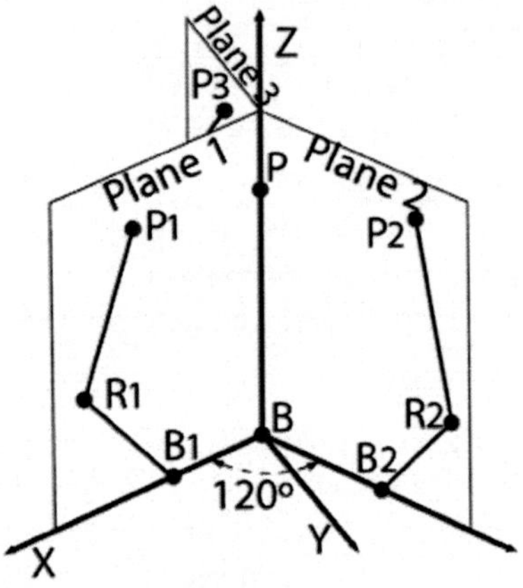

Fig. 3 Proposed vertical planes

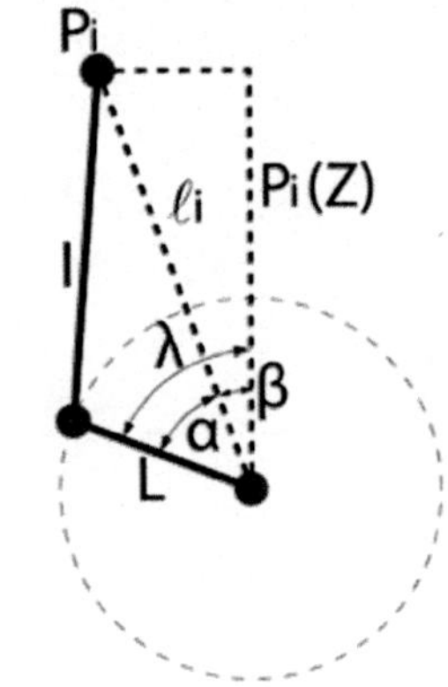

Fig. 4 Kinematic legs geometry

$$\alpha_i = \arccos \frac{L^2 + \ell_i^2 - l^2}{2\,L\ell_i} \tag{8}$$

$$\beta_i = \arccos \frac{P_{dzi}}{\ell_i} \tag{9}$$

$$\lambda_i = \alpha_i + \beta_i \tag{10}$$

3.2 Forward Kinematics

The problem of forward kinematics is more complex than the inverse kinematics, therefore a simulation based solution is presented, in this approach the position of the

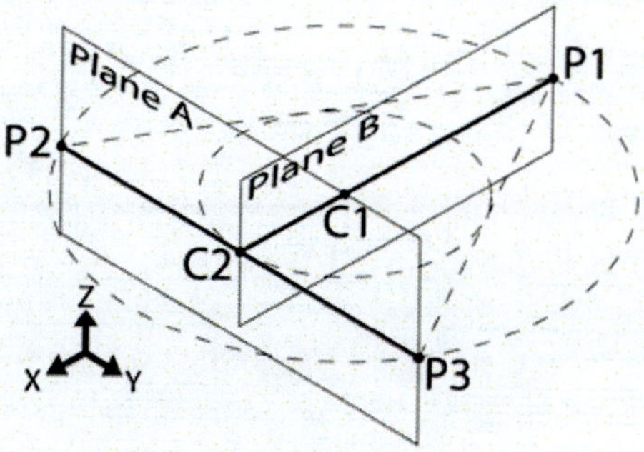

Fig. 5 Proposed geometry of the platform using points P_i

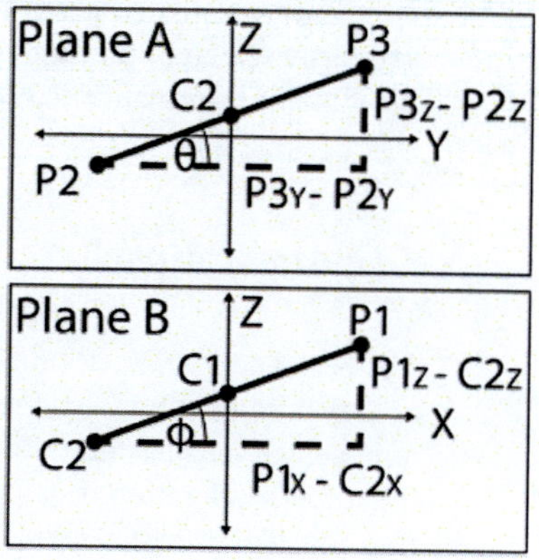

Fig. 6 frontal view of the planes generated from P_i

union points of the platform (P_{ri}) are assumed to be known, the proposed geometry for the platform its shown in Fig. 5, the Planes A and B defined in Fig. 5 are shown in Fig. 6, points $P2$ and $P3$, defined in plane (A), are used to calculate the platform roll, then (11) is used to calculate the pitch of the platform (ϕ); (12) to calculate point $C2$ on plane A; (13) together with $C2$ on plane B, to calculate the pitch of platform (θ); and (14) is used to calculate heave (ΔZ).

$$\phi = \arctan \frac{P3_Z - P2_Z}{P3_Y - P2_Y} \tag{11}$$

$$C2 = \frac{P2_Z + P3_Z}{2} \tag{12}$$

$$\theta = \arctan \frac{P1_Z - C2_Z}{P1_x - C2_x} \tag{13}$$

$$m = \frac{P1_X - C2_X}{P1_Z - C2_Z}$$
$$\Delta Z = P1_Z - mP1_X \tag{14}$$

4 Control Design

As regards control, the selected strategy must be capable of performing position tracking of a reference signal and compensating for external disturbances, on the other hand, the system must have a saturation that maintains the system within the workspace, in this research the maximum an minimum position of the actuator is arbitrarily selected since the scope of this project is not related with the determination of optimal performance to the workspace of the 3-RRS parallel manipulator.

4.1 Motor Model

The permanent magnet, brushed, DC gear-motor can be modeled in many ways, e.g. as a system of the first or second order, nevertheless, in this research the motor is represented as a perturbed double integrator, and (15) is proposed. In this equation the rotor acceleration is calculated as a function of the armature voltage and it takes into consideration the endogen perturbations as the factor including angular velocity, and the exogen perturbations as the factor that containing the torque applied to the motor. Once the motor's acceleration is calculated, it is integrated two times to obtain the position.

$$\ddot{q} = \left(\frac{K_a K_b}{J_m R} + \frac{B_m}{J_m}\right)\dot{q} + \frac{K_a}{J_m r R}V - \frac{\tau}{J_m r^2} \tag{15}$$

where:

- q = angular position.
- $\dot{q}$ = angular velocity.
- $\ddot{q}$ = angular acceleration.
- K_a = motor-torque constant.
- K_b = counter electromotive constant.
- J_m = rotor inertia.
- R = armature resistance.
- B_m = viscous resistance.
- τ = torque applied to the motor.
- r = gear relation.

4.2 ADRC Control

The selected control strategy is the active disturbance rejection control (ADRC) since it has the feature of compensating internal and external disturbances. This control strategy was first introduced in [8], the main principle behind this strategy is to reduce all perturbations to a single term that is compensated in the control law. To estimate the disturbance, an extended state observer (ESO) is implemented, this observer estimates the states of the system and the perturbation based on the control signal and the output of the system, the methodology to determine the observability, controllability and stability of the system are beyond the scope of this research, Fig. 7 shows the block diagram of the ADRC.

To design the control, (15) is modified to group all disturbances in a single term ζ, the resulting expression is shown in (16), then, making $\ddot{q} = u$, where u is the control signal, and leaving only v on the left side of the equation as shown in (17); defining the system states as (18), the next step is to define a control law, the proposed control law is stated in (19) where x_d represents the motor's desired position.

$$\ddot{q} = \frac{K_a}{J_m r R} V + \zeta \tag{16}$$

$$V = \frac{J_m r R}{K_a}(u - \zeta) \tag{17}$$

$$\begin{aligned} x_1 &= q \qquad \dot{x_1} = x_2 \\ x_2 &= \dot{q} \qquad \dot{x_2} = u + \zeta \end{aligned} \tag{18}$$

$$u = \ddot{x_d} - K_1(x_1 - x_d) - K_2(x_2 - \dot{x_d}) - z_1 \tag{19}$$

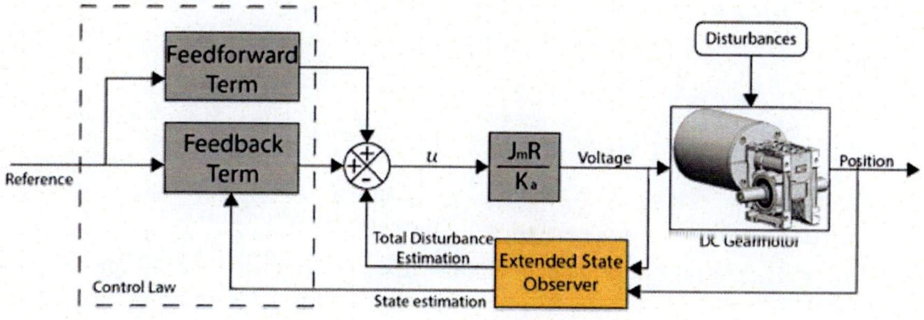

Fig. 7 ADRC block diagram based on [9]

where:

- K_i = control gains $i = (1, 2)$
- z_1 = estimated disturbance ($z_1 \approx \zeta$)

4.2.1 Extended State Observer

The ESO is a key feature of the ADRC since it can estimate the system disturbance z_1, and the states of the system $\hat{x_i}$ with $i = (1, 2)$, the equations that define the observer are shown in (20), these equations use the constants o_i with $(i = (0 - 3))$, these constant values must be carefully calculated since they rule over the speed of the observer, as a rule of thumb, the ESO must be faster than the controlled system but the exact values needed to tune the system require a fair amount of trial and error, nevertheless there is a useful methodology to calculate those values, a fourth-order Hurwitz polynomial is defined as shown in (21). Then the roots of that equation are used as the constants o_i. This leaves only the natural frequency ω_n and the damping factor ξ to be defined as shown in (22).

$$\begin{aligned} \dot{\hat{x_1}} &= \hat{x_2} + o_3(x_1 - \hat{x_1}) \\ \dot{\hat{x_2}} &= u + z_1 + o_2(x_1 - \hat{x_1}) \\ \dot{z_1} &= z_2 + o_1(x_1 - \hat{x_1}) \\ \dot{z_2} &= o_0(x_1 - \hat{x_1}) \end{aligned} \tag{20}$$

$$(s^2 + 2\omega_n \xi s + \omega_n^2)(s^2 + 2\omega_n \xi s + \omega_n^2) = 0 \tag{21}$$

$$\begin{aligned} o_3 &= 4\omega_n \xi \\ o_2 &= 2\omega_n^2(1 + 2\xi^2) \\ o_1 &= 4\omega_n^3 \xi \\ o_0 &= \omega_n^4 \end{aligned} \tag{22}$$

5 Simulation Results

The simulation was performed using MATLAB/Simulink, the basis of the simulation is a 3D platform that uses the Simscape complement, the 3D model is shown in Fig. 8 in the block called "platform", in this model it is possible to visualize the platform's movement, measure the position of points P_i and measure the position and acceleration of points B_i. Those latter points are actuated using the output of the control strategy.

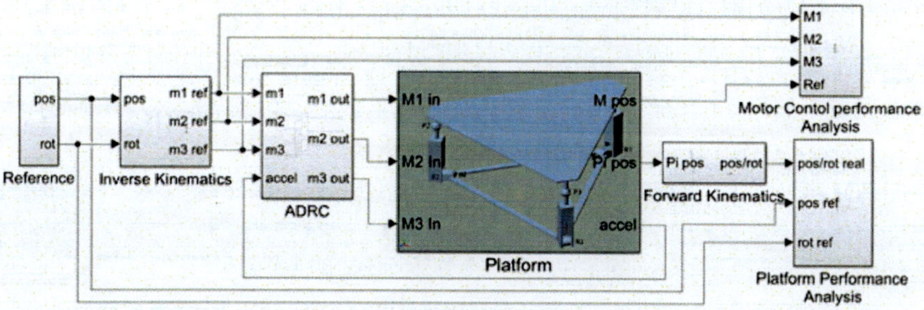

Fig. 8 MATLAB/Simulink simulation

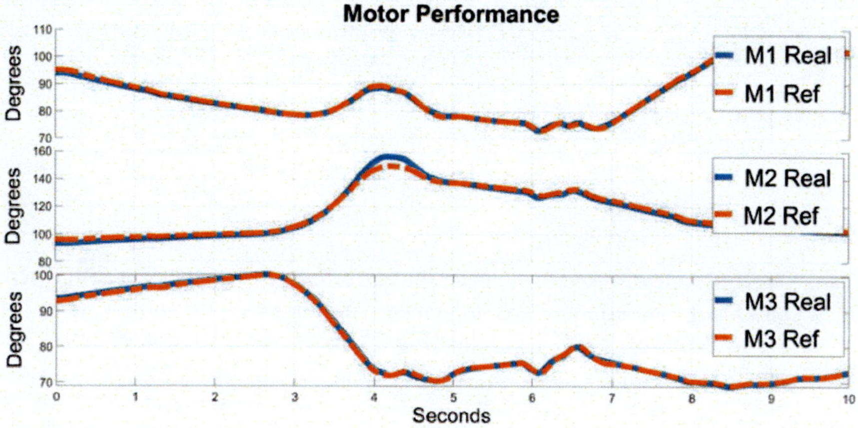

Fig. 9 DC motor position control performance analysis

The overall structure of the simulation is shown in Fig. 8, the reference position of the platform is being generated from an Excel sheet that contains the reference position of a driving simulator, the nature of this signal is beyond the scope of this research, nevertheless this signal is being used to measure the performance of the platform.

The control strategy was simulated using the motor model, the ADRC control and the ESO; at the same time the measurement of the acceleration in the joints is compared with the ideal acceleration of the motor and that difference is added as a disturbance, the performance of the control is shown in Fig. 9 where the plotted position is the measured position of the joints and the reference is the reference generated from the inverse kinematics.

The overall performance of the system is being measured comparing the input reference position (roll, pitch and heave) with the measured position of the platform, Fig. 10 shows the comparison between these signals. To visualize the performance of the system quantitatively, the performance index ISE (integral squared error) is calculated.

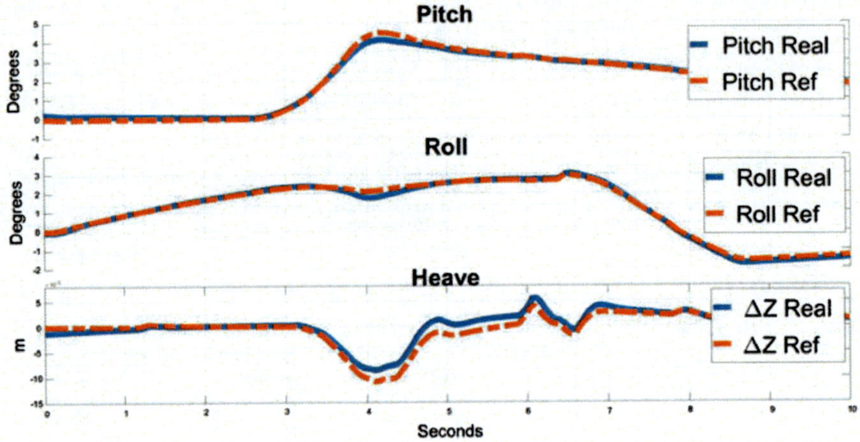

Fig. 10 Platform performance analysis

6 Discussion

The ISE index is used to compare the performance of the system with other proposals. The proposed system is compared with the ISE index of the system changing the ADRC with a PID controller, Fig. 11 shows the evolution of both indexes.. To obtain the PID performance, the controller was tuned with the PID Tuner provided by MATLAB/Simulink and the disturbance of the system was multiplied by several orders of magnitude with the goal of simulating a loaded platform. Therefore, while the inverse kinematics also plays an important role in the performance of the platform, the main source of error comes from the control strategy and the disturbances affecting the system.

Regarding the mathematical model of the platform, the RRS configuration, compared with the RP configuration, exchanges workspace for speed, also, given the non-linearity of the system, for the model to remain relevant, the angle (λ) must not get near 0° or 180°, where the model reaches a point where it fails completely, therefore, maintaining the system within its worspace is vital.

7 Conclusion

The inverse kinematics of a 3-RRS parallel manipulator is presented alongside the modeling of the actuators and the design of an ADRC for position tracking. The system was simulated with MATLAB/Simulink and simscape. The presented mathematical model of a 3-RRS parallel manipulator can follow the reference as long as the angle λ remains approximately between 10° and 170°.

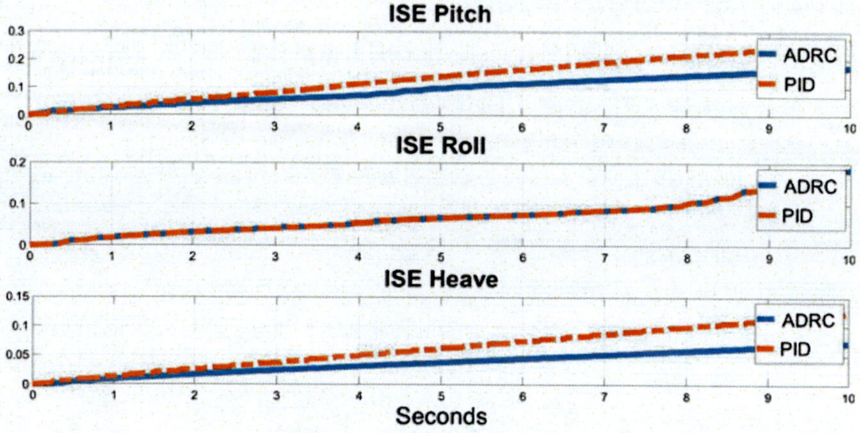

Fig. 11 Platform's performance ISE Index comparison

The simulated system can track the given reference and compensate the intrinsic and extrinsic disturbances caused by the platform weight, inertia, etc. The results obtained in this work are valuable references for the design and implementation of physical driving simulators. Further work for this model needs to be experimentally tested and implemented to measure the performance of the system.

References

1. Slob JJ (2008) State-of-the-art driving simulators, a literature survey. DCT Report. https://doi.org/10.10007/1234567890
2. Fisher DL, Rizzo M, Caird JK, Lee JD (2011) Handbook of driving simulation for engineering, medicine, and psychology. CRC Press
3. Brecht D (2015) A 3-DOF Stewart platform for trenchless pipeline rehabilitation. Sept 2015, pp 1–185
4. Bürüncük K, Tokad Y (1999) On the Kinematic of a 3-DOF Stewart platform. J Robot Syst 16(2):105–118
5. Itul T, Pisla D (2009) On the kinematics and dynamics of 3-DOF parallel robots with triangle platform. J Vibroeng 11(1):188–200
6. Li J, Wang J, Chou W, Zhang Y, Wang T, Zhang Q (2001) Inverse kinematics and dynamics of the 3-RRS parallel platform. In: IEEE International Conference on Robotics and Automation, vol. 3, pp 2506–2511. https://doi.org/10.1109/ROBOT.2001.932999
7. Carbajal-Espinosa O, Izar-Bonilla F, Díaz-Rodríguez M, Bayro-Corrochano E (2016) Inverse kinematics of a 3 DOF parallel manipulator: a conformal geometric algebra approach. In: IEEE-RAS international conference on humanoid robots, Dec 2016, pp 766–771. https://doi.org/10.1109/HUMANOIDS.2016.7803360

8. Huang H, Wu L, Han J, Feng G, Lin Y (2004) A new synthesis method for unit coordinated control system in thermal power plant—ADRC control scheme. In: 2004 International conference on power system technology, POWERCON 2004, vol 1, Nov 2004, pp 133–138. https://doi.org/10.1109/icpst.2004.1459980
9. Guerrero-Castellanos JF, Rifaï H, Arnez-Paniagua V, Linares-Flores J, Saynes-Torres L, Mohammed S (2018) Robust active disturbance rejection control via control Lyapunov functions: application to actuated-ankle-foot-orthosis. Control Eng Pract 80:49–60. https://doi.org/10.1016/j.conengprac.2018.08.008

Study of a Denatured Bovine Serum Albumin Solution Used as Lubricant in Tribological Testing of Total Knee Replacements

G. I. Girón de la Cruz, J. D. O. Barceinas-Sánchez, and M. Gómez-Ramírez

Abstract The lubricant solution used in tribological testing of total knee replacements must have a 20 g/L (ISO 14243-3:2014) overall value of protein concentration; bovine serums (BSs) are currently used as base fluids to prepare such solution. BSs normally have a higher concentration of bovine serum albumin (BSA) and minor concentration of globulins (α, β and γ). It has been observed that these proteins can be degraded and adsorbed onto the surface of the materials during tribological tests. Therefore, this research is aimed at analyzing the proteins of denatured lubricant BSA-based solutions having a protein concentration of 20 g/L, after 6 h of tribological testing under different conditions. The testing tribological parameters considered were the applied load (L), sliding-to-rolling ratio (SRR), and entrainment speed (V_m). Four combination of parameters were set: (a) 8N-18-80 mm/s, (b) 2N-18-80 mm/s, (c) 8N-0.2-20 mm/s, and (d) 2N-0.2-20 mm/s. All tests carried out at 37 °C. Samples of the solutions were analyzed before and after testing by UV–Vis spectroscopy (range 200–300 nm), Bradford and Ellman methods. The UV–Vis spectroscopy analyzes allowed to determine possible change suffered of the aromatic residues (residues exposed or covered cause an increase or decrease of absorbance). Bradford analysis of solutions indicated similar content of proteins. Therefore, basic amino acid residues in proteins interact and reorganize. Moreover, disulfide bonds involved in tertiary structure of protein remains stable. Finally, COF (Coefficient friction) is dependent on both, protein state as well as the tribological parameters.

Keywords Bovine serum albumin · Tribological testing · Denaturation

G. I. Girón de la Cruz · J. D. O. Barceinas-Sánchez (✉) · M. Gómez-Ramírez (✉)
Instituto Politécnico Nacional—Centro de Investigación en Ciencia Aplicada y Tecnología Avanzada Unidad Querétaro, Cerro Blanco No. 141, 76090 Querétaro, QRO, México
e-mail: obarceinas@ipn.mx

M. Gómez-Ramírez
e-mail: mgomezr@ipn.mx

K. L. Flores Rodríguez et al. (eds.), *Recent Trends in Sustainable Engineering*, Lecture Notes in Networks and Systems 297,
https://doi.org/10.1007/978-3-030-82064-0_7

1 Introduction

The lubricant solution used in tribological testing of total knee replacements must have a 20 g/L (ISO 14243-3:2014) overall value of protein concentration [8]; Bovine serums are currently used as base fluids to prepare the lubricant solution, particularly fetal bovine serum (FBS) has a higher concentration of bovine serum albumin (BSA) and a minor concentration of globulins (α, β and γ). It has been observed that these proteins degrade and adsorb onto the surface of the materials during the tests, moreover there is interest on how every compound works, particularly the denatured BSA [4, 23].

FBS has a total protein concentration that has a minimal variation according to suppliers, that in the case of BIOWEST, it contains a total of 36 g/L, composed by 16 g/L of BSA, 14.1 g/L of α-globulin, 5.5 g/L of β-globulin and 0.5 g/L of γ-globulin (Catalog N°: S1650-500). It is useful, at this stage, to examine the features of proteins in more detail. They consist of high molecular weight polypeptides formed by amino acid with a typical structural hierarchy (from primary to quaternary structure). Primary structure is given by peptidic bonding, secondary structure by H-bonding (α helix and the β pleated sheet). For tertiary structure also H-bonding, hydrophobic interaction, ion bonding and disulfide bonds (three-dimensional arrangement) [23].

BSA is a protein with a molecular weight of 66.8 kDa and it has low levels of methionine and tryptophan, and a high level of cystine and charged amino acids like glutamic, aspartic, arginine and lysine. Glycine and isoleucine have lower BSA levels than average proteins. It is consisting of 582 amino acids and the structure is stabilized by the cross-linked 17 disulfide bridges of cysteine amino acids. Its sequence also contains a free cysteine at position 34 [10–12].

Therefore, in the present paper was analyzed protein changes of denatured lubricant BSA-based solutions at 20 g/L, after 6 h of tribological testing with different parameters. Samples were analyzed before and the end of tribological testing by UV–Vis spectroscopy (200–300 nm), the Bradford and Ellman methods, to determine possible change suffered by the proteins in solution at the end of tribological tests.

2 Materials and Methods

2.1 Preparation of Bovine Serum Albumin Solutions (Denatured)

Commercially available BSA lyophilized powder, ($\geq$96% purity) has been purchased from Sigma–Aldrich. The BSA was soluble under distilled water with protein concentration of 20 g/L. Thermal denaturation solution: This was achieved by heating the BSA during incubation at 70 °C for 30 min, then it was subjected to tribological testing [18]. Moreover, before those solutions were subjected to testing, they were

Table 1 Parameters (fixed) in the tribometer

Fixed parameters
Profile type: Ball on disc (3/4 ball)
Step type: Timed step
Temperature: 37 °C
Step duration: 6:00:00 (hh:mm:ss)
Logging Interval: 180 s (121 points)

analyzed (0 h ctrl-denatured), too. Also, it was prepared control solution that only was incubated at 37 °C for 6 h (37 °C-6 h ctrl-denatured) that is do not subjected to tribological testing.

2.2 *Test Conditions for Tribological Test*

In this study was used an MTM2 ball-on-disc tribometer (PCS Instruments, London, UK). The steel ball, filled against the surface of the UHMWPE (ultra-high molecular weight polyethylene) disc, is driven separately of this one. This consists of a AISI316L stainless steelball (19.05 mm diameter) and a UHMWPE disc (46 mm diameter and 6.00 ± 0.02 mm thickness) [1, 7].

The parameters for the tribological tests were "applied load (L)"; "sliding-to-rolling ratio (SRR)", and "entrainment speed (V_m)". Four combination parameters were using (always in the same order): (a) L = 8N-SRR = 18-V_m = 80 mm/s, (b) L = 2N-SRR = 18-V_m = 80 mm/s, (c) L = 8N-SRR = 0.2-V_m = 20 mm/s and (d) L = 2N-SRR = 0.2-V_m = 20 mm/s. The fixed parameters programmed in the tribometer are shown in Table 1. The coefficient of friction (COF) was an acquired response.

2.3 *Experimental Techniques*

2.3.1 UV–Vis Spectroscopy

For present study, a scan from 200 to 300 nm was performed to identify the absorption spectrum that allowed visualizing the variation of the absorbance presented by the lubricant solution to identify conformational transitions. In general, differences in absorbance between forms of a protein (the native and the denatured) are small, but highly useful for monitoring a protein's conformational changes [21].

2.3.2 Bradford Method

The present study analyzes the changes in proteins concentration by microassays Bradford method [13], solutions were prepared by diluting at 0.025 g/L. It was used 800 μL of sample diluted and 200 μL Bradford reagent (B6916-Sigma Aldrich). Absorbance was measured immediately at room temperature at 595 nm against the sample and reagent blanks. The total protein concentration in solution was determined according to a BSA calibration curve (2–40 μg/mL). Values were expressed in g/L for all solutions [25].

2.3.3 Ellman Method

The reagent DTNB (D8130-Sigma-Aldrich) stock solution 2 mM, was prepared in sodium acetate 50 Mm by using molecular biology grade water. Tris solution (1 M adjusted the pH to 8.0) was prepared. Finally mixing 100 μl of Tris solution, 50 μl of the DTNB stock solution, and 840 μl of molecular grade biology H_2O. The final volume was 1000 μl when 10 μl of sample was added. Absorbance was measured at 412 nm. The concentration of free sulfhydryl groups in solution was calculated using a L-cysteine-monohydrate calibration curve (1–10 mmol/L). Values were expressed in mmol/L for all solutions [5].

3 Results

3.1 UV–Vis Spectroscopy

A molecule may exhibit different behavior of absorption wavelength and strength of absorbance depending on how it chromophores are in its chemical nature and its molecular environment. Absorption spectroscopy is thus a great way to follow ligand-binding reactions, conformational transitions and enzyme catalysis in proteins [21]. The principle it is about absorption of radiation in the near UV by proteins depends on the Tryptophane (Trp) and Tyrosine (Tyr) content (and to a very small extent on the amount of Phenylalanine (Phe).

Particularly in BSA (PDB ID:3V03) has following amounts of the aromatic amino acids and characteristics: Trp (2 residues, maximum absorption peak at 280 nm and absorptivity of 5600); Tyr (20 residues, maximum absorption peak at 274 nm and absorptivity of 1400) and Phe (27 residues, maximum absorption peak at 257 nm and absorptivity of 200) (https://www.ks.uiuc.edu/Research/vmd) [22]. In native proteins due that the aromatic residues are buried (hydrophobic core) it has seen that show a slight increase in wavelength, otherwise they are exposed in the denatured proteins showing a slight decrease in wavelength. Tests were performed to determine changes in proteins (aromatic amino acids covered or exposed) at the end of tribological assays for the solution shows a signal in the UV–Vis spectrophotometer were made dilutions until final concentration was 0.125 g/L.

Figure 1 shows the absorption spectrum (200–300 nm) of the denatured BSA-based solutions in control system and solutions subject to tribological parameters,

Table 2 shows the wavelengths (217 and 279 nm) where were detected two maximum absorbance values (λmax). The solution measured at 0 h (ctrl-native) showed 1.344 a.u. (red line), this solution was only BSA undenatured and do not subjected to tribological test, nor to 37 °C during 6 h. About the solution measured at 0 h (ctrl-denatured) showed 1.252 a.u. (black line), this solution was only BSA denatured and do not subjected to tribological test, the reduction in absorption suggest covering aromatic residues due denaturation. Moreover, the incubated denatured BSA-based solution (37 °C–6 h) showed a value of absorbance of 1.341 a.u. (black dash line) (do not subjected to tribological test), which suggest that during incubation time at 37 °C there could be a reorganization of the denatured protein causing

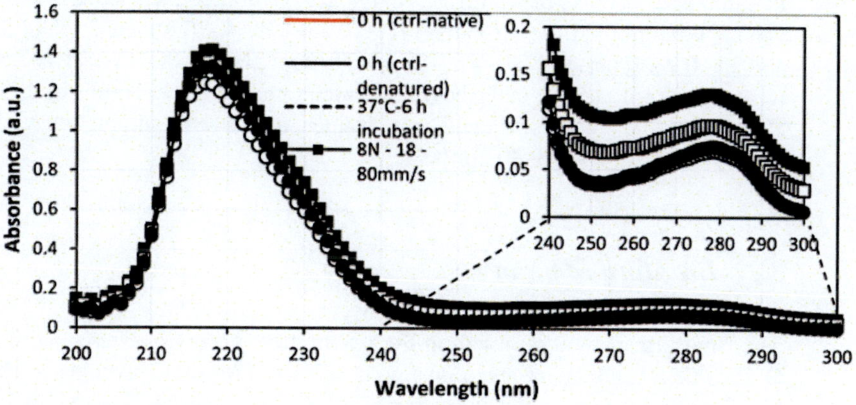

Fig. 1 Absorption spectrum of the denatured BSA-based solutions using tribological parameters

Table 2 Absorbances obtained at 217 and 279 nm in all solutions

	First λmax values (217 nm)	Second λmax values (279 nm)
Test	Absorbance (a.u.)	Absorbance (a.u.)
0 h (ctrl-ative)	1.344	0.076
0 h (ctrl-denatured)	1.252	0.067
37 °C-6 h incubation	1.341	0.072
8N-18-80 mm/s	1.403	0.129
2N-18-80 mm/s	1.287	0.096
8N-0.2-20 mm/s	1.239	0.069
2N-0.2-0 mm/s	1.318	0.074

exposure of some aromatic residues causing the absorbance was similarly to BSA native.

In the other hand, it was showed the absorbance in tribological test after 6 h, at 37 °C. Test with high values: L = 8N-SRR = 18-Vm = 80 mm/s (black square) the absorbance at 217 nm was 1.403 a.u. and shows an increase compared with T = 0 h ctrl-native, T = 0 h ctrl-denatured and BSA denatured incubating at 37 °C–6 h, which suggest exposed aromatic residues due parameters used. Then it was showed the test with parameters L = 2N-SRR = 18-Vm = 80 mm/s (white square), the absorbance was 1.287 a.u. (like ctrl-denatured). For test with parameters L = 8N-SRR = 0.2-Vm = 20 mm/s showed 1.239 a.u. (white circle) like ctrl-denatured, too. And the last test was realized with low parameters: L = 2N-SRR = 0.2-Vm = 20 mm/s, the absorbance was 1.318 a.u. (black circle), this suggests that changes in the conformation of proteins because aromatic residues were exposed. In general, changes in absorbance could be due, to increase or decrease of exposition of the aromatic residues contained in the BSA solution by conditions used in tribological assays. In summary it is possible to identify three groups with similar trend of absorbance values: First group: 0 h (ctrl-native) (1.344 a.u.), 37 °C–6 h incubation (1.341 a.u.) and 2N–0.2–20 mm/s (1.318 a.u.). Second group: 2N–18–80 mm/s (1.287 a.u.), 0 h (ctrl-denatured) (1.252) and 8N–0.2–20 mm/s (1.239 a.u.). Third group: 8N–18–80 mm/s (1.403 a.u.). For second λmax values, is showed a similar behavior except to absorbance of 2N–18–80 mm/s.

3.2 *Bradford Method*

The Bradford method is based on the reaction between Coomassie Brilliant Blue G-250 (CBB) and the basic amino acid in proteins (Arginine, Lysine and Histidine) [3]. There is an electrostatic interaction of the CBB species binding to proteins, since it involves negative charges of CBB and positive charged amino acid side chains. Particularly, basic amino acids in BSA (PDB ID:3V03) has following amounts of the basic amino acids: Arg (23 residues); Lys (59 residues) and His (17 residues) (https://www.ks.uiuc.edu/Research/vmd) [9, 14].

Test were performed to quantify protein total content and additionally to know the availability of basic amino acids due conformational change. Dilutions were made until final concentration of 0.025 g/L before analysis. Values were expressed in g/L for all solutions. The bars in figures represent the standard deviation of the means. It is observed BSA-based solutions using all tribological parameters determined previously.

Figure 2 shows the protein content of all solutions using tribological parameters. Although there is no noticeable change, with respect to the controls, there is a tendency to decrease the protein concentration in most of the solutions in contrast to 0 h (ctrl-native) and 0 h (ctrl-denatured) but there are not significant differences (Table 3). This suggests a decrease availability of basic amino acids and thus a decrease the protein–dye binding (basic residues were exposed) showing a slight increase to

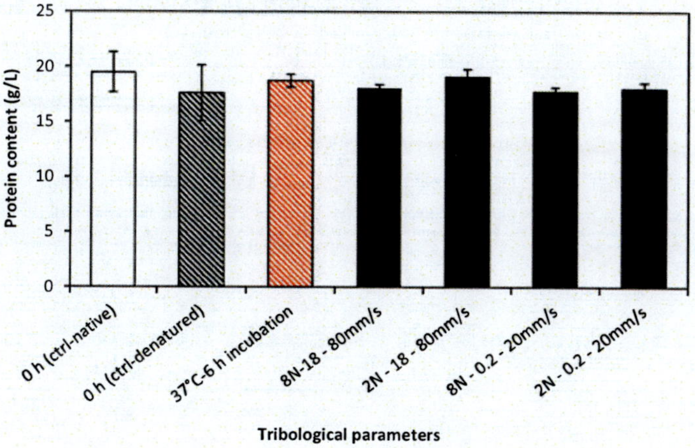

Fig. 2 Protein content in tribological tests of the denatured BSA-based solutions

Table 3 Tukey's test (group comparisons for One-Way ANOVA) (Minitab Statistical Software)

Factor	N	Mean	Group
0 h (ctrl-native)	5	19.487	A
2N-18-80 mm/s	5	19.095	A
37 °C-6 h incubation	5	18.742	A
8N-18-80 mm/s	5	18.095	A
2N-0.2-20 mm/s	5	18.058	A
8N-0.2-20 mm/s	5	17.756	A
0 h (ctrl-denatured)	5	17.59	A

content protein. Moreover, this suggests that BSA is more susceptible to change and covered basic amino acids.

3.3 Ellman Method

Ellman's reagent, 5, 5′-dithio-bis-(2-nitrobenzoic acid), also known as DTNB, is compound that consist about determination of free sulfhydryl groups. A mixed disulfide is formed when DTNB reacts with a free sulfhydryl group releasing a yellow colored product (2-nitro-5-thiobenzoic acid known as NTB), which is measurable at 412 nm. Particularly, the protein BSA shows 34 residues of cysteine and 17 disulfide bonds [10]. In solutions of BSA, free sulfhydryl groups content has been established spectrophotometrically in accordance with Ellman's method. Test were performed to quantitating free sulfhydryl groups in solution. Dilutions were made until final

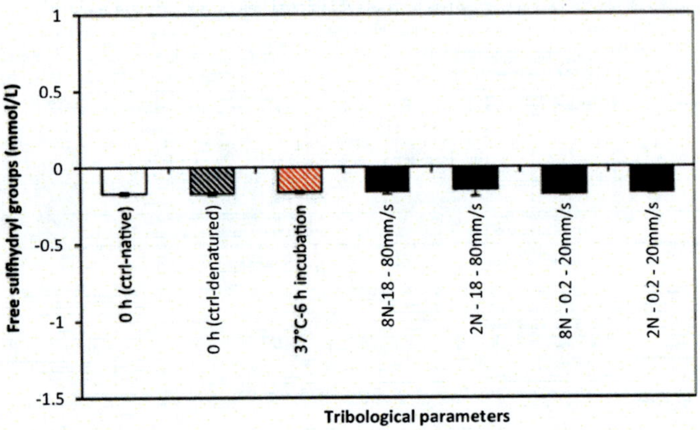

Fig. 3 Determination of free sulfhydryl groups in tribological test of denatured BSA-based solutions

concentration was 1 g/L before to apply Ellman method. Values were expressed in mmol/L for all solutions.

The bars in figures represent the standard deviation of the means. The results for denatured BSA-based solutions obtained with Ellman method are shown using four combination of tribological parameters. Only negative amounts of free sulfhydryl groups were obtained, suggesting that no cleavage of the disulfide bonds present in the proteins. Therefore, the stability of the tertiary structure of proteins in all solutions is not compromised by any of the tribological parameters used (Fig. 3).

3.4 *Behavior of the Coefficient of Friction*

Additionally, the variation of the COF over timed up to 6 h (21,600 s) is shown in Fig. 4. It is observed denatured BSA-based solutions using tribological parameters determined previously. For the parameters L = 8N-SRR = 18-Vm = 80 mm/s (high values) (black line) yielded initial and final COF values of 0.045 and 0.47, respectively. About the tribological parameters (when L changed to lower value): L = 2N-SRR = 18-Vm = 80 mm/s (blue line) the COF behavior was completely different (initial COF was 0.077 and final 0.0003) without further significant decay along time.

However, about the tribological parameters L = 8N-SRR = 0.2-Vm = 20 mm/s (gray line) COF does follow a similar tendency to high values along time but showed an initial higher COF (0.112) and final lower value (0.0003). Both do not decay along time. About parameters (when L changed to higher value) L = 2N-SRR = 0.2-Vm = 20 mm/s (low values) (red line), the COF showed a different behavior like that of

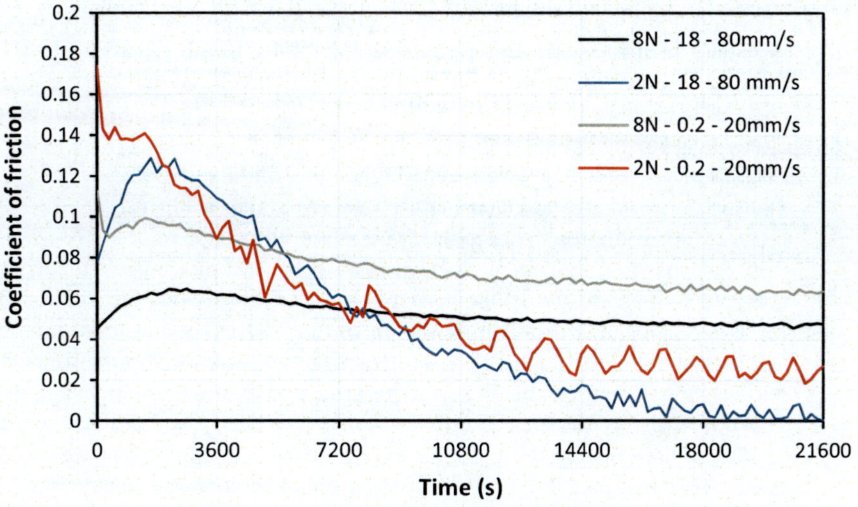

Fig. 4 Coefficient of friction of tribological test using denatured BSA-based solutions

L = 2N-SRR = 18-Vm = 80 mm/s). The COF for the lower tribological parameters showed initial and final values of 0.174 and 0.026, respectively.

In summary, for tribological tests applying the higher load, the COF showed a similar tendency and different initial COF value. This could be related to the effect of protein denaturation. About this tribological test applying the lower load, COF showed similar tendency however, the decay time and the fluctuations were different. It could be possible that the denatured BSA produced such fluctuations.

4 Discussion

Since the absorbance values that informs about denatured BSA do not correspond to lubricant solutions subjected to testing tribological however, it is not possible found reported data to compare against the results generated in the present study [20]. Alternatively, it has been reported conformational changes of proteins for researchers interested on the artificial joint lubricating mechanisms through Cobalt Chromium Molybdenum (CoCrMo) ball-on-flat wear tests whit human synovial fluid samples and 25 wt% bovine calf serum. Post-test organic surface deposits were studied by Absorption Spectroscopy and their results reported the deposits were particularly denatured proteins with higher content of β-sheet.

Then, to identify denatured β-sheet structures it was used fluorescent imaging of deposited films by Thioflavin T molecular dye and their results indicated aggregated denatured β-sheet fibrils. About deposits presence in the wear scar (by Scanning Electron Miscroscopy), their results showed a reduced scar due deposited protein.

Therefore they suggest that the generation of not soluble, denatured protein films is supposed the main lubrication mechanism facilitated the protection of surface during rubbing [23]. In comparison with the present work that has shown the conformational changes of proteins in denatured BSA-based solutions, particularly, was observed the effect on aromatic amino acids (Tyr, Trp and Phe) [21] with absorbances differences in relation to tribological test. Due to higher absorptivity of Trp, the results indicates that in BSA solutions affected mainly two Trp amino acid (residues 134 and 212) [10–12]. Also, the tests tribological parameter with high values (8N-18-80 mm/s) (Fig. 2) caused more evident changes in absorbances, this suggest that tribological parameters increase the BSA damage. However, absorbance in solution that was incubated (37 °C-6 h incubation) was similar to the 0 h ctrl-native solution, suggesting that it could be possible there were interactions between BSA proteins, and for that reason, the solution be able to maintain stable and do not present changes.

About total protein concentration, studies have been identified that reliant on the content of both BSA and γ-globulin shown changed in COF. It has been studied that there were a maximal level of BSA content and in another way an physiological level of γ-globulin content for them to make effective lubricants [4]. Other researchers elaborated the method for lubricant film generation and it was reported that the layer formed of albumin is thicker in comparison to layer γ-globulin [15]. In other study, the presence of γ-globulin and hyaluronic acid significantly improved cartilage friction only when the osteoarthritis was in advanced-stage [17]. In another study, significant differences were found between untested and test serum fetal deposits that indicated protein denaturation since the deposits reported a higher denatured β-sheet content. This was observed through a test was developed following reciprocating sliding conditions to detect the friction and wear of the hip implant materials. Their results suggest that fluid volumes minor than 1.5 ml were influenced by loss through evaporation consequently increasing the protein content and triggering a lower wear. Denatured proteins were found in the deposits so that analysis verified the importance of serum fetal proteins in relation to determination of wear in CoCrMo pair [24]. The studies mentioned shown the importance not only of proteins (particularly BSA and γ-globulin) but also other components in FBS. Moreover, it is important to know the stability of each proteins (BSA and globulins) in a complex FBS solution. This does not coincide with our results, because the denatured BSA-based solutions used not shown significant changes in protein concentration. These results suggest that there was a slightly decrease in the availability of basic amino acids (Arg, Lys and His), probably due proteins were previously denatured and which results to conformational change (hiding amino acids that could interact whit CBB).

About the stability of the tertiary structure of proteins in all solutions is not compromised by any of the tribological parameters used, because the free sulfhydryl groups have not been exposed. However, typical results on free sulfhydryl groups have been extensively involved as important functional groups in many dietary proteins (i.e., Ellman's reagent used in egg white, flour and milk (also their products) [2].

Exposure of sulfhydryl groups (in the processing of ulra-high-temperature (UHT-9s exposure of 151.7 °C of sterilized milk) in whey proteins occurs concomitantly with the development of cooked flavor, and, in fact, appearance of this flavor seems to be dependent upon their presence. Instability of whey proteins during heat treatment (skim milk heated to 100 °C or 90 °C for 30 min) than also appears to be associated with changes in the sulfhydryl/disulfide groups of these proteins [19]. For the above reason, the present study confirms that tribological conditions and previous BSA denaturation at do not be able to exposure of sulfhydryl groups like treatment mentioned (heat treatment).

The behavior COF dependent of the experimental parameters: Vm, SRR, and L. However, it has been observed that COF is minor when the lubricating fluid is diluted and is used along the whole walking cycle. Additionally, researchers have proposed that |Vm| had a slightly effect on the COF for testing fluids [7]. Other researchers showed a correlation (linear) between the thickness of the film (adsorbed denatured protein) and the COF when was used a low sliding speed (10 mm/s). Otherwise, proteins could maintained stable their secondary structure at major sliding speed (50 mm/s), but different behavior of albumin and γ-globulin proteins could be observed was respect to its different structure [16]. In another study, the researchers found that heat processes caused the secondary structure of albumin to be lost. The denatured structure leads to a major adsorption rate onto material surface due to its compactness which results in the increase of COF. They suggest that a compact layer of denatured albumin protein is formed and tends to be adsorbed to the hydrophobic UHMWPE surface which may trigger in the increase of COF [6].

According to present study, it shows that is important to monitor the state of the protein (if it is denatured or not) in lubricant solutions before and then performing the tribological tests, since it is observed that it influences the behavior of the COF. In the other hand, the variation in four combination parameters, there was different COF behavior. However, COF observed differences seem to be related to tribological parameters, as well as changes in relation to BSA denatured state.

5 Conclusions

This work analyzed the changes suffered by the proteins in solution previously denatured at the end of tribological assays. Findings are summarized below:

1. The changes in the absorbances detected in the BSA after tribological assays could be due to increase or decrease of exposition of the aromatic residues contained in the protein.
2. Results of Bradford analysis of solutions indicate that the solutions had similar content of proteins. No significant differences were observed between different combination of tribological parameters. Therefore, basic amino acid residues in proteins interact and reorganize each other, even when subjected to denaturing and tribological testing.
3. During tribological assays by using denatured BSA the stronger interaction due to disulfide bonds involved in tertiary structure of protein remains stable.

4. It has been demonstrated that COF is dependent on both, state protein (native and denatured) and tribological conditions.

Acknowledgements The authors gratefully acknowledge the support from the Instituto Politécnico Nacional, SIP Project 20195708, and the National Council for Science and Technology (CONACYT).

References

1. Barceinas-Sanchez JDO, Alvarez-Vera M, Montoya-Santiyanes LA, Dominguez-Lopez I, Garcia-Garcia AL (2017) The coefficient of friction of UHMWPE along an entire walking cycle using a ball-on-disc tribometer under arthrokinematics and loading conditions prescribed by ISO 14243–3:2014. J Mech Behav Biomed Mater 65:274–280. https://doi.org/10.1016/j.jmbbm.2016.08.032
2. Beveridge T, Toma SJ, Nakai S (1974) Determination of SH- and SS- groups in some food proteins using Ellman's Reagent. J Food Sci 39(1):49–51. https://doi.org/10.1111/j.1365-2621.1974.tb00984.x
3. Bradford MM (1976) A rapid and sensitive method for the quantitation of microgram quantities of protein utilizing the principle of protein-dye binding. Anal Biochem 72(1–2):248–254. https://doi.org/10.1016/0003-2697(76)90527-3
4. Duong CT, Lee JH, Cho Y, Nam JS, Kim HN, Lee SS, Park S (2012) Effect of protein concentrations of bovine serum albumin and γ-globulin on the frictional response of a cobalt-chromium femoral head. J Mater Sci Mater Med 23(5):1323–1330. https://doi.org/10.1007/s10856-012-4603-9
5. Ellman G (2018) DTNB (Ellman's Reagent) | GoldBio. Goldbio.com from https://www.goldbio.com/product/1199/dtnbellmans-reagent
6. Fang HW, Hsieh MC, Huang HT, Tsai CY, Chang MH (2009) Conformational and adsorptive characteristics of albumin affect interfacial protein boundary lubrication: from experimental to molecular dynamics simulation approaches. Colloids Surf B 68(2):171–177. https://doi.org/10.1016/j.colsurfb.2008.09.029
7. Garcia-Garcia AL, Alvarez-Vera M, Montoya-Santiyanes LA, Dominguez-Lopez I, Montes-Seguedo JL, Sosa-Savedra JC, Barceinas-Sanchez JDOO (2018) Regression models to predict the behavior of the coefficient of friction of AISI 316L on UHMWPE under ISO 14243–3 conditions. J Mech Behav Biomed Mater 82:248–256. https://doi.org/10.1016/j.jmbbm.2018.03.028
8. ISO 14243-3 (2014) Implants for surgery—wear of total knee-joint prostheses—Part 3: loading and displacement parameters for wear-testing machines with dis-placement control and corresponding environmental conditions for test, Second edition. International Organization
9. Georgiou CD, Grintzalis K, Zervoudakis G, Papapostolou I (2008) Mechanism of Coomassie brilliant blue G-250 binding to proteins: a hydrophobic assay for nanogram quantities of proteins. Anal Bioanal Chem 391(1):391–403. https://doi.org/10.1007/s00216-008-1996-x
10. Jahanban-Esfahlan A, Ostadrahimi A, Jahanban-Esfahlan R, Roufegarinejad L, Tabibiazar M, Amarowicz R (2019) Recent developments in the detection of bovine serum albumin. Int J Biol Macromol 138:602–617. https://doi.org/10.1016/j.ijbiomac.2019.07.096
11. Jahanban-Esfahlan A, Panahi-Azar V (2016) Interaction of glutathione with bovine serum albumin: spectroscopy and molecular docking HO. Food Chem 202:426–431. https://doi.org/10.1016/j.foodchem.2016.02.026
12. Jahanban-Esfahlan A, Panahi-Azar V, Sajedi S (2015) Spectroscopic and molecular docking studies on the interaction between N-acetyl cysteine and bovine serum albumin. Biopolymers 103(11):638–645. https://doi.org/10.1002/bip.22697

13. Kruger NJ (1994) The bradford method for protein quantitation. In: Walker JM (eds) Basic protein and peptide protocols. Methods in molecular biology™, vol 32. Humana Press. https://doi.org/10.1385/0-89603-268-X:9
14. Ku HK, Lim HM, Oh KH, Yang HJ, Jeong JS, Kim SK (2013) Interpretation of protein quantitation using the Bradford assay: comparison with two calculation models. Anal Biochem 434(1):178–180. https://doi.org/10.1016/j.ab.2012.10.045
15. Nečas D, Sadecká K, Vrbka M, Gallo J, Galandáková A, Křupka I, Hartl M (2019) Observation of lubrication mechanisms in knee replacement: a pilot study. Biotribology 17:1–7. https://doi.org/10.1016/j.biotri.2019.02.001
16. Nečas D, Sawae Y, Fujisawa T, Nakashima K, Morita T, Yamaguchi T, Vrbka M, Křupka I, Hartl M (2017) The influence of proteins and speed on friction and adsorption of metal/UHMWPE contact pair. Biotribology 11:51–59. https://doi.org/10.1016/j.biotri.2017.03.003
17. Park J, Duong C, Sharma A, Son K, Thompson M, Park S et al (2014) Effects of hyaluronic acid and γ–globulin concentrations on the frictional response of human osteoarthritic articular cartilage. Plos ONE 9(11):e112684. https://doi.org/10.1371/journal.pone.0112684
18. Park JH, Jackman JA, Ferhan AR, Ma GJ, Yoon BK, Cho NJ (2018) Temperature-induced denaturation of bsa protein molecules for improved surface passivation coatings. ACS Appl Mater Interfaces 10(38):32047–32057. https://doi.org/10.1021/acsami.8b13749
19. Patrick PS, Swaisgood HE (1976) Sulfhydryl and disulfide groups in skim milk as affected by direct ultra-high-temperature heating and subsequent storage. J Dairy Sci 59(4):594–600. https://doi.org/10.3168/jds.S0022-0302(76)84246-4
20. Pignataro MF, Herrera MG, Dodero VI (2020) Evaluation of peptide/protein self-assembly and aggregation by spectroscopic methods. Molecules 25:4854
21. Schmid F (2001) Biological macromolecules: UV-visible spectrophotometry. Encycl Life Sci. https://doi.org/10.1038/npg.els.0003142
22. Skoog D, Holler F, Crouch S, Skoog D (2008) Principios de análisis instrumental. 6e. Cengage Learning Editores S.A. de C.V
23. Stevenson H, Jaggard M, Akhbari P, Vaghela U, Gupte C, Cann P (2019) The role of denatured synovial fluid proteins in the lubrication of artificial joints. Biotribology 17:49–63. https://doi.org/10.1016/j.biotri.2019.03.003
24. Stevenson H, Parkes M, Austin L, Jaggard M, Akhbari, P, Vaghela U et al (2018) The development of a smallscale wear test for CoCrMo specimens with human synovial fluid. Biotribology 14:1–10. https://doi.org/10.1016/j.biotri.2018.04.001
25. Walker J (2002) The protein protocols handbook. 2e. Humana Press. https://doi.org/10.1385/1592591698

Screening of Pectinolytic Activity and Bioconversion of Ferulic Acid to Aromatic Compounds from *B. cereus* IFVB and *B. subtilis* IFVB Isolated Mexican Vanilla (*Vanilla planifolia* ex. Andrews) Beans from the Curing Process

Esmeralda Escobar-Muciño, Margarita M. P. Arenas-Hernández, and Ma. Lorena Luna-Guevara

Abstract This investigation aimed to evaluate the pectinolytic activity and the bioconversion of ferulic acid (FA) to aromatic compounds (CAS) in *Bacillus sp.* strains isolated from curing Mexican *Vanilla planifolia.* The isolated (30) from 3 different stages and 3 regions of the vanilla bean curing process. The pectinolytic activity and bioconversion were evaluated of isolated by the cup-plate method with MM-Pectin agar (0.5%) and MM-Vanilla (0.01% of FA) agar, respectively. The results indicated that *Bacillus* strains degraded pectin and assimilated FA. Nine strains of *Bacillus cereus* and *Bacillus subtilis* exhibited the higher pectinolytic activity index (PAI) (4.54 ± 0.07 and 3.15 ± 1.66) and bioconversion of ferulic acid index (BFAI) (4.5 ± 0.05 and 4.40 ± 0.07) were obtained of reception stage from Papantla 3 region. While strains 19 and 27 of *B. subtilis* presented BFAI values of 3.55 ± 0.16 and 3.56 ± 0.33 isolated to killing stage from Papantla 2 and Ayotoxco regions. A total of nine strains of *B. cereus* and *B. subtilis* showed PAI (4.23 ± 0.38) and IBAF (4.44 ± 0.4) values from selection stage and the mentioned regions. Finally, the *Bacillus* species had the capacity of degrade pectin and the potential for bioconversion AF to CAS. Denoting *B. subtilis* which showed its highest activity in the stages 1 and 2 and *B. cereus* presented it particularly in stage 3 of the vanilla curing process. This is the first report about quantitative plate assay of pectinolytic activity and evaluation of the capacity of bioconversion the FA to CAS by the agar plate from *B. cereus* and *B. subtilis* isolated from Mexican *V. planifolia* beans.

E. Escobar-Muciño · M. M. P. Arenas-Hernández (✉)
Centro de Investigación en Ciencias Microbiológicas, Posgrado en Microbiología, Instituto de Ciencias, Benemérita Universidad Autónoma de Puebla, Puebla, Puebla, México
e-mail: margarita.arenas@correo.buap.mx

Ma. L. Luna-Guevara (✉)
Colegío de Ingeniería en Alimentos, Facultad de Ingeniería Química, Benemérita Universidad Autónoma de Puebla, Puebla, Puebla, México
e-mail: maria.luna@correo.buap.mx

K. L. Flores Rodríguez et al. (eds.), *Recent Trends in Sustainable Engineering*, Lecture Notes in Networks and Systems 297,
https://doi.org/10.1007/978-3-030-82064-0_8

Keywords *Bacillus cereus* · *Bacillus subtilis* · Pectinolytic activity · Bioconversion · Bioconversion to aromatic compounds · Stages of vanilla curing · Ferulic acid production regions · Vanilla

1 Introduction

Natural vanilla is derived from the fruits of the *Vanilla planifolia* orchid, and it is the second most valuable flavoring in the food processing industry. The aroma and taste are developed in vanilla beans during the curing process, which occurs after harvest. Traditional curing vanilla is produced in México as well as some other countries such as Madagascar and Indonesia. This process differs in each country, which consists of several stages: (I) selection of vanilla beans, (II) killing, (III) sun curing and successive sweating, (IV) classification of vanilla beans, (V) conditioning or storage of vanilla beans according to size, color, flexibility, and aroma [1, 2].

Ferulic acid is among the main precursors of the aromatic compounds generated during the curing of vanilla. It has been reported that natural vanillin can be produced through microbial conversion from FA which is found in the plant cell wall. This production process is low cost compared to the biosynthetic process [3].

The Totonacapan region grown vanilla, and it has a traditional curing process, this allows the color change of fruit from green to dark brown, increasing the vanillin concentration vanilla (20,000 ppm) by enzymatic activity in the cell vegetal wall [2, 4]. The Mexican curing process is handcraft and lasts between 3 and 6 months, the vanilla beans are exposed to the sun. During this process, the different microorganisms have been isolated in vanilla beans, and the most frequent are some species of enterobacteria [5]. Until now, the behavior of *Bacillus* under a vanilla curing process in México is unknown.

Other studies describe different microorganisms isolated from vanilla beans under a curing system, in Indonesia, the isolates were obtained from the first and last stage of the process. Among these isolates, *Bacillus* (*B. subtilis*, *B. licheniformis* and *B. smithii*) was found as the main microorganism isolated from green beans. Moreover, the genus *Bacillus* has been reported with hydrolytic enzymatic activity on the plant cell wall, suggesting that bacterial enzymatic contributes to the production of the phenolic compound "vanillin" by releasing certain precursors [1]. *Bacillus* can produce a beneficial effect on vanilla beans by the production of lipopeptides, whose facilitate the colonization of cured bean and the leaf of *V. planifolia* plant [1, 4, 6]. Besides, there is a contribution to the curing process of the vanilla bean by *Bacillus* favoring the increase in the production of volatile components as vanillic acid, which contributes to the sensory attributes of vanilla beans [4].

Pectin degrading bacterial strains have been reported through drop-plate and cup-plate methods using pure colony and quantitative colorimetric technique using bacterial supernatant [7, 8]. Demonstrating that both techniques are useful to find pectin degrading strains, but the plate methodology stands out, which is a rapid screening method [8–10]. In addition, the presence of pectinolytic enzymes have been studied in some bacteria through o bioinformatic techniques and 3 main families were reported [11, 12].

On the other hand, the Gram-positive bacteria as *Bacillus* can be endophytic and produce pectinolytic enzymes. These enzymes hydrolyze the plant cell wall and releasing substrates or precursors that are used in different enzymatic reactions for the bioconversion. The FA bioconverts to vanillin and subsequently, when it is oxidized for produces the vanillic acid, which is finally demethylated to protocatechuate [13, 14]. Also, it has been suggested that *Bacillus* may participates in the second stage of vanilla curing (or killing) for release of vanillin precursors [1].

Some strains from the genus *Bacillus* have been reported to be tolerant to vanillin because they possess enzymes that bioconvert FA to aromatic compounds [5, 6, 11, 12]. This characteristic is a process of adaptation of the microorganisms in response to the high concentrations of vanilla that are toxic in some microorganisms. For this reason, the bacterial have developed cellular detoxification mechanisms, which enables the rapid degradation of aromatic compounds as vanillin and vanillic acid, producing other degradation products, and finally obtaining acetyl-CoA, which is recycled and used by the Krebs cycle [6, 11]. Also, direct bacterial conversion of FA to vanillin has been evidenced in *Bacillus* and it is suggested that the release of FA by the action of bacterial feruloyl esterase enzyme contributes to the induction of pectinolytic enzymes [2, 5]. The present work aimed to evaluate the association between the pectinolytic activity and the bioconversion of FA to aromatic compounds of *Bacillus* strains and the stages and sites of the vanilla curing process.

2 Materials and Methods

2.1 Vegetal Material

Vanilla beans (*V. planifolia* ex. Andrews) were obtained from 3 different stages of the curing process (reception, killing of the vegetal cell wall and selection of quality vanilla cured beans). The fruits were collected from various locations in the Totonacapan region as Ayotoxco, Pantepec and Papantla (latitude 20°, 23′, 35.6″ and longitude 97°, 19′, 34.72″) in Puebla and Veracruz, México. The vanilla beans were sampled from November 2016 to April 2017, and they were stored at 4 °C in polystyrene bags for later use.

2.2 Isolation and Identification of **Bacillus** *sp. from the Surface of Vanilla Fruits*

A total of 60 isolates were obtained from *Vanilla planifolia* beans in 3 different stages of curing process (reception, killing of the vegetal cell wall, and selection of quality

Table 1 Methods and conditions of study to evaluate the processes involved in aromatic compounds production by *B. cereus* and *B. subtilis* bacteria isolated from vanilla beans

Process	Method	Types of sample	Measurement parameter
Pectinolytic activity (PA)	Cup-plate	Supernatant	PAI
Bioconversion of ferulic acid (BFA)	Cup-plate	Pellet	BFAI

Diameter and PA index. *BFAI* Bioconversion of ferulic acid index. The PAI and BFAI are dimensionless

vanilla cured beans). The colonial morphology identification of *Bacillus cereus* and *Bacillus subtilis* was realized on a selective and differential MYP medium (Mannitol-Egg yolk and Polymyxin) [15]. The standard *B. cereus* colonies were rough and dry, with a bright pink backdrop and an egg yolk precipitate surrounding of colony and *B. subtilis* colonies changed the color of the orange to yellow [16]. The selective colonies were grown in TSB (Tryptic Soy Broth) and were stored in 1.5 mL tubes containing 30% glycerol at −80 °C, they were used for the pectinolytic activity and the FA bioconversion screenings [17]. The methodologies included in this analysis, as well as the types of samples used and the required parameters to evaluate the bacterial processes involved in aromatic compound production, are described in Table 1.

2.3 *Preparation of Samples: Pellets and Supernatant*

From cryopreservation vial (10 μL) of Gram-positive bacteria was inoculated into a volume of 10 mL TSB and incubated for 24 h at 37 °C. The bacterial culture obtained was centrifuged for 10 min at 13,000 g, and the pellet was resuspended in 150 μL of Tris–HCl buffer (50 mM, pH 7.5). And the supernatant was recovered and precipitated with methanol (1:2 v/v), before the methanolic extract was centrifuged at 13,000 g for 10 min and the pellet obtained was resuspended in the same buffer until they are used for the pectinolytic activity and the FA bioconversion screenings.

2.4 *Evaluation of the Processes Involved in the Production of Aromatic Compounds*

The Pectinolytic Activity (PA) was measured using the cup-plate method from samples of protein extracts obtained from supernatant and the Pectinolytic Activity Index (PAI) was reported. The same method was used to evaluate the ferulic acid assimilation from the pellet, and the Bioconversion of Ferulic Acid Index (BFAI) was determined.

2.4.1 Pectinolytic Activity (PA)

With the help of a cork borer, a cup of 9 mm in diameter was made in the M9 minimal medium. Each cup was inoculated with 150 μL of bacterial culture (supernatant or pellet) and incubated at 37 and 50 °C for four days. The plates were covered with 10 mL of Congo red overnight, subsequently, 10 mL of NaCl solution (5 M) was used for removing the dye for 1–3 days until clearing zones around the cups were observed. The diameter of the clearing zone was considered as the criteria for determining the PA of the bacterial strain according to Hasegawa et al. [18]. The pectinolytic activity index (PAI) was calculated by Eq. 1:

$$\text{PAI} = \frac{\text{Diameter of hydrolysis zone}}{\text{Diameter cup}} \tag{1}$$

2.4.2 Ferulic Acid Assimilation

To determine the FA assimilation in the Gram-positive bacteria, 20 mL of bacterial culture was centrifuged at 19,300 g for 10 min and the pellet was washed and resuspended in 50 mM of Tris–HCl buffer, pH 7.0. The agar plates on minimal vanillin-producing medium (MVPM), contained 1.5% agar supplemented with 0.1% of FA (w/v) and 2% of glucose (w/v). A cylindrical cavity was made in the medium with a sterile cork borer. The cavity was filled with 150 μL of each sample [18, 19]. The MVPM plates were incubated at 37 °C for 12–24 h and a color change was observed (green to yellow), indicating the assimilation of FA and glucose as carbon source [19]. The diameter of the yellow halo around the cup was measured for each sample and Eq. 2 was used to calculate the Bioconversion of Ferulic Acid Index (BFAI) in all samples by the cup-plate method:

$$\text{BFAI} = \frac{\text{Diameter of hydrolysis zone}}{\text{Diameter cup}} \tag{2}$$

2.5 Data Analysis

Samples (pellet and supernatant) and Gram-positive bacteria (*B. cereus* IFVB and *B. subtilis* IFVB) were considered as variation resources. The results were analyzed using a design experiment equivalent to the completely randomized balanced. The comparison of means between the treatments was carried out by the Tukey test ($p \leq 0.05$). Statistical analysis was performed with the Statistix 10 software and reported results are the means of the duplicates in 2 independent experiments.

3 Results

3.1 *Isolation and Selection of* Bacillus *Strains from the Mexican Vanilla Curing Process*

The colonial morphology was considered to two species: *B. cereus* IFVB (isolation frequency 39%), which was characterized by a defined colony with pink coloration in the selective medium, and *B. subtilis* IFVB (isolation frequency 61%), which is characterized by a defined colony with yellow coloration in the selective medium. The 67 Gram-positive bacteria were obtained from investigated curing stages, as well as the three producer regions of curing vanilla (Ayotoxco, Pantepec, and Papantla). It was found that Gram-positive bacteria could be isolated from 1, 2, and 4 stages of the Mexican vanilla bean curing process. Table 2 showed the results of *Bacillus* isolates from different regions of the vanilla bean curing process in the MYP selective medium.

3.2 *Pectinolytic Activity in* Bacillus *Isolates*

The results obtained in this study showed that the 30 *Bacillus* IFVB strains hydrolyzed the pectin (*B. subtilis* 21 strains and *B. cereus* IFVB 9 strains). Besides, *B. cereus* strain (7) IFVB isolated from the Papantla-3 region reached the highest value of PAI (4.54 $\pm$ 0.07) with the cup-plate method. This microorganism demonstrated the strongest biotechnological potential for hydrolyzing pectin in the stage 1 (reception stage of vanilla beans) under in vitro conditions. While the results in the stage 2 (killing of the vegetal cell wall) were positive 12 strains of *B. subtilis* IFVB for the pectinolytic activity test. The highest PAI (3.15 $\pm$ 0.07) was obtained from *B. subtilis* strain (14) IFVB isolated from the Papantla-2 region. This strain showed the important biotechnological potential for hydrolyzing pectin in the stage 2 (killing stage of vanilla) beans. In the third stage of vanilla curing beans, no Gram-positive bacteria were isolated (no data).

The results of the pectinolytic activity test on 9 *Bacillus* strains (6 strains of *B. cereus* IFVB and 3 strains of *B. subtilis* IFVB) were positive in the fourth stage of vanilla curing beans (selection of high-quality cured vanilla beans). The high PAI (4.23 $\pm$ 0.38) value was reached from *B. cereus* strain (63) IFVB isolated from the Ayotoxco region. Both strains exhibited biotechnological potential for hydrolyzing pectin and releasing aromatic precursors during the fourth stage of the vanilla curing process. Figure 1a–c illustrates the results of the formation of pectinolytic activity halos in plate, while Fig. 1d shows the PAI quantified from *Bacillus* isolates using the cup-plate method.

Table 2 *Bacillus* isolates in the MYP selective medium from different regions of the Mexican vanilla curing process

Stage of curing	Region of vanilla curing	Isolated species	Number of isolates
1 (reception of vanilla beans)	Ayotoxco, Puebla	*B. subtilis* IFVB	3
	Pantepec, Puebla	*B. subtilis* IFVB	2
	Papantla 1, Veracruz	*B. subtilis* IFVB	1
	Papantla 2, Veracruz	*B. subtilis* IFVB	1
	Papantla 3, Veracruz	*B. subtilis* IFVB	2
		B. cereus IFVB	2
2 (killing of the vegetal cell wall of vanilla beans)	Ayotoxco, Puebla	*B. cereus* IFVB	2
		B. subtilis IFVB	5
	Pantepec, Puebla	*B. cereus* IFVB	1
		B. subtilis IFVB	2
	Papantla 1, Veracruz	*B. cereus* IFVB	2
		B. subtilis IFVB	1
		Bacillus sp. IFVB	9
	Papantla 2, Veracruz	*B. cereus* IFVB	1
		B. subtilis IFVB	5
	Papantla 3, Veracruz	*B. cereus* IFVB	1
4 (selection of quality vanilla curing beans)		*B. cereus* IFVB	3
	Pantepec, Puebla	*B. cereus* IFVB	6
		B. subtilis IFVB	3
	Papantla 1, Veracruz	*B. cereus* IFVB	2
	Papantla 2, Veracruz	*B. cereus* IFVB	1
		B. subtilis IFVB	6
	Papantla 3, Veracruz	*B. cereus* IFVB	1
		B. subtilis IFVB	5

The acronym IFVB indicates an isolated from vanilla beans

3.3 Ferulic Acid Bioconversion in the Genus Bacillus

The findings demonstrated that nine strains were positive in the screening of FA assimilation in the first stage of vanilla curing beans (*B. cereus* IFVB with 3 strains and *B. subtilis* IFVB with 6 strains). The strains isolated from the Papantla-3 region showed a higher BFAI (4.5 ± 0.05), suggesting a considerable participation to contribute to the aromatic profile of vanilla beans in comparison with other strains of the stage 1. The screening of FA assimilation suggested 12 positive *B. subtilis* IFVB strains in the second stage of vanilla curing beans (killing of the vegetal cell wall). *B. subtilis* strain (157) from the Papantla 2 region presented a BFAI of 4.44 ± 0.40 confirming its contribution to the aromatic profile of vanilla beans. Finally,

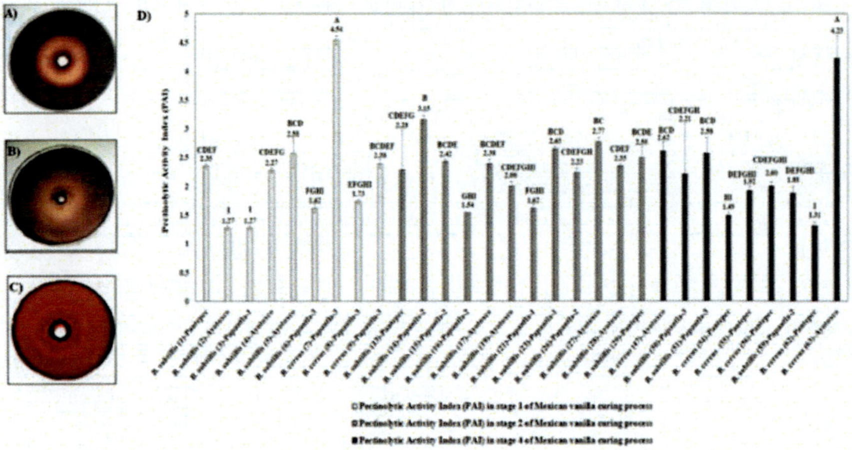

Fig. 1 Pectinolytic activity in *B. cereus* and *B. subtilis* was determined by the cup-plate method (**a**–**c**) using enzymatic extract obtained of strains isolated from stage 1, 2 and 4 of the curing processes of vanilla beans on the MM-P agar. Furthermore, the pectinolytic activity index (PAI) of *Bacillus* strains was obtained using the cup-plate method (**d**). Equal capital letters indicate that there was no significant difference between the samples according to the Tukey test ($P \geq 0.05$)

the results obtained in the fourth stage of vanilla curing beans were 9 strains positive (*B. cereus* IFVB with 6 strains and *B. subtilis* IFVB with 3 strains). *B. subtilis* strain (62) from the Pantepec region exhibit an value of BFAI 2.33 ± 0.40, indicating an intermediately biotechnological potential for contribution to the aromatic profile of vanilla beans in comparison with other results of the stage 4. The formation of Ferulic Acid Assimilation (FAA) halos on the MVPM by the cup-plate are observed in Fig. 2a–c. And the Bioconversion of Ferulic Acid Index (BFAI) data of *Bacillus* isolates using the cup-plate method are showed in Fig. 2d.

Finally, MM-VP agar plates with *Bacillus* strains after 24 h of incubating, aromas were detectable in the petri dishes, indicating the formation of aromatic compounds. With these observations, it can be believed that the isolated microorganisms can contribute to the release of precursors through the action of pectinolytic activity.

4 Discussion

4.1 Isolation and Selection of Bacillus *Strains from the 3 Stages of Vanilla Bean Curing*

Bacillus sp., *B. subtilis*, and *B. cereus* have been isolated from several fruits and have been shown to degrade pectin through plate methodology [1, 10, 20–24]. The genus *Bacillus* is reported as a microorganism capable of contributing to the biosynthesis of

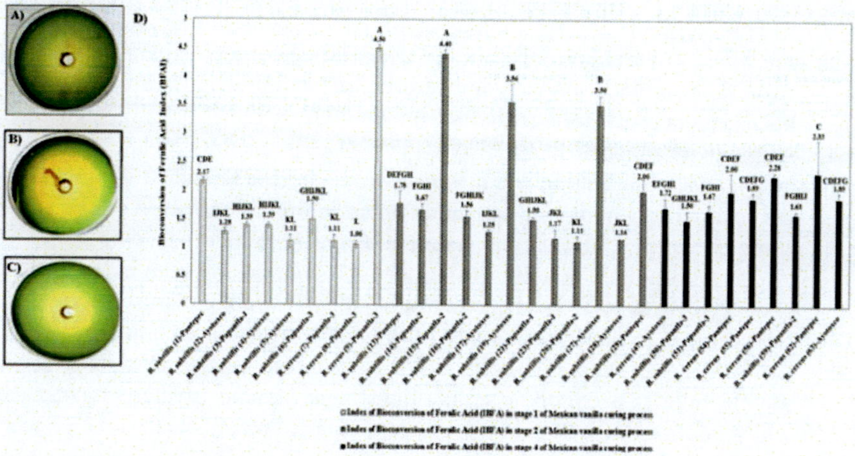

Fig. 2 Ferulic acid assimilation (FAA) in *B. cereus* and *B. subtilis* was obtained by the cup-plate method using bacterial pellet from isolated stages 1, 2 and 4 of vanilla bean curing process in MVPM agar plate (**a–c**). The bioconversion of ferulic acid index (BFAI) from *B. cereus* IFVB and *B. subtilis* IFVB (**d**). Equal capital letters indicate that there was no significant difference between the samples according to the Tukey test ($P \geq 0.05$)

aromatic compounds in vanilla bean fruits as the vanillin [25]. One of the most widely studied microorganisms is *B. vanillea* sp. isolated from vanilla beans cured process involved in the production of vanilla aroma compounds [4, 13]. On the other hand, *B. subtilis* has been reported as a frequent isolate during the vanilla bean curing process steps and *s* there are few reports about importance of *B. cereus* in the production of the aromatic compounds. However, some species of *Bacillus* have been utilized by improving the production of aromatic compounds and even influence the growth of plants and the control of the stem and root diseases of *V. planifolia* [1, 6].

4.2 *Pectinolytic Activity*

The studies by Röling et al. [1] found differences in the abundance of microorganisms and the characteristics that influence the production of aromatic compounds. The variations in the aroma of vanilla beans by the Gram-positive bacteria like the genus *Bacillus* are related to six pectinolytic enzymes that participate in the hydrolysis of the plant cell wall, which are divided into the 3 main families [1, 12, 26]. The *Bacillus* enzyme are polygalacturonases that act on the non-reducing of pectin, hydrolyze the methyl group of pectin (pectin methylesterase), and hydrolyze pectin oligomers to tetrasaccharides and disaccharides (pectate lyase), according to bioinformatic study [12].

Few studies of enzymatic plate hydrolysis have been reported in *B. subtilis* and *B. cereus*, which showed that these bacterial species no producing constitutive pectinases. The current research found the premise for optimizing hydrolysis conditions on plate, a higher pectinolyic activity from *Bacillus sp.* can be determined in comparison with previous reports like Pickaver [27]. Also, in the present study the pectinolytic activity was evaluated in two different temperatures (37 and 50 °C), confirming that the optimal for the PA in the genus *Bacillus* was 37 °C being consistent with the study of Soares et al. [28] and Kaur and Gupta [29].

The pectinolytic activity in the genus *Bacillus* using pure colony and enzyme extract by plate assay has been measured, demonstrating that *B. subtilis* hydrolyzes the pectin as in the present study [30]. Other reports have confirmed that the plate method test can be used to evaluate a large number of bacterial isolates, allowing the identification of pectinolytic strains, including a *Bacillus* isolated from decayed fruits and vegetables [20]. In another research, 72 *Bacillus* strains from fruits were tested using the same method, showing that the majority of the strains hydrolyzed pectin. This suggests that bacterial fruit isolates have high pectinolytic activity [10], as seen in the strains *B. cereus* and *B. subtilis* isolates from beans of vanilla in the current study. An analysis made with 26 bacterial genera revealed that *B. subtilis* developed hydrolase activity by plate test. These results show that certain bacteria develop different characteristics to adapt to availability of nutrient and that the environment has an impact on gene rearrangement and the reorganization of genetic information to activate the production of hydrolytic enzymes constitutive. These bacterial enzymes are essential for the breakdown of complex polymeric substrates such as pectin from fruits of vanilla beans [1, 23, 31]. This finding was observed with *B. cereus* IFVB and *B. subtilis* IFVB that presented a high pectinolytic activity. A higher percentage of *Bacillus* strains have been reported positive for pectinolytic activity associated to the hydrolysis clear zones, that were comparable with the results of this study [32].

Different studies have demonstrated that *Bacillus* can degrade pectin using a plate test, with diameters of 0.5–1.5 cm like zone of substrate hydrolysis of [28, 33]. In contrast with this research, we found that several strains produced higher hydrolysis diameters, indicating that *B. cereus* and *B. subtilis* are considered as strains that produce a high pectinolytic activity using plate test. Furthermore, our results were consistent with Varghese et al. study [9], they tested the pectin hydrolysis in thirteen isolates from fruits. This research group found that *B. subtilis* was the best producer of pectinase reporting the maximum zone of hydrolysis enzymatic by plate assay [8, 9]. These results were confirmed by Kavuthodi et al. [34] whom indicated that *B. subtilis* strain BKDS1 exhibited larger hydrolysis of pectin area (2.6 cm) by cup-plate assay, which is comparable with the diameters obtained from *B. cereus* IFVB and *B. subtilis* IFVB in the present research [34]. Other authors reported that two *Bacillus* isolates hydrolyzed pectin with hydrolysis zones of 4 and 4.5 cm by the cup-plate method [35]. The same report mentioned that *B. cereus* isolated from fruits has the potential to deteriorate pectin agar plate incubated at 37 °C for 24 h The obtained results on PA and the clear hydrolysis region in the plates, which were close to values observed in other researches, concluding that the best condition of incubation in *B. cereus* IFVB and *B. subtilis* was 37 °C [10, 21, 22]. Besides, Husseiny et al.

[36] reported that it was possible to identify the hydrolysis of pectin in *Bacillus* employing the cup-plate method, evaluating the production of pectinolytic enzymes, but the diameters and PA index were not reported. Finally, that few studies have been published about quantitative plate assay of pectinolytic activity in *B. cereus* and *B. subtilis* [36].

4.3 The Bioconversion of FA to Aromatic Compounds

The synergy between the microbial enzymatic activity in the plant cell wall and the aroma precursors as FA, has been confirmed [4]. Also, the results of this study consider the genus *Bacillus* participation in the Mexican vanilla curing process, due to bacteria was isolated from the killing stage of vanilla curing. Also, this genus *Bacillus* is resistant to the high temperatures [1] therefore it was isolated the second stage of vanilla curing process. Assimilation of FA from *B. cereus* and *B. subtilis* was confirmed in this study, as source carbon in a time range of 2–24 h, observing that from 2 h of incubation the bioconversion produced yellow halos being consistent with other studies [19, 31]. And was observed that the yellow halos of bioconversion of FA are more intense in 24 h of monitoring, being comparable with the reports by Chen et al. [3] and Zamzuri et al. [19] whose investigated the assimilation of the glucovainillin and ferulic acid in an equivalent time of bioconversion [19]. Alike other metabolites such as vanillic acid, vanillin, the protocatechuic acid, among other aromatic compounds are produced from FA by *Bacillus* [4]. Also, *B. vanillea* and *B. subtilis* B7-S produce aromatic compounds from the FA [4, 14, 37], suggesting that the *B. subtilis* IFVB isolated from vanilla beans could bioconvert the FA to aromatic compounds, nonetheless there are few reports about the bioconversion process in *B. cereus*. It has been characterized the enzyme ferulic acid decarboxylase from *B. cereus* strain SAS-3006, whose biological role in the bioconvertion of FA to vanillin and other aromatic compounds [38, 39].

An interesting fact is that the genus *Bacillus* presented the ability to produce pectinolytic enzymes as was demonstrated in the present study, also can assimilate FA and bioconvert to aromatic compounds during the 3 stages of the vanilla bean curing process. The above showed a correlation between both processes demonstrating the participation of *Bacillus* in the hydrolysis of pectin in the cell wall vegetal and the bioconversion of FA to aromatic compounds, which were confirmed in the plate's experiments [1, 11]. Otherwise, in this study was observed that *B. cereus* IFVB and *B. subtilis* IFVB incubated in agar MVPM Petri dishes produce vanillin-like aromas, which is an indicative of the production of aromatic compounds under in vitro conditions, confirmed by Gu et al. [25] demonstrated the possible participation of the genus *Bacillus* to add sensory attributes to vanilla beans. Moreover, *B. cereus* and *B. subtilis* produces enzymes that participate in the bioconversion of substrates like FA to aromatic compounds, and the study of these enzymes results important to understanding the biological role of the *Bacillus* genus. There are few reports about the sequence data related to the genes and enzymes responsible to produce vanillin, such as the feruloyl esterase, vanillin oxygen oxidoreductase (HcaB), the vanillate O-demethylase oxygenase A and B (VanA/VanB) have been reported in a few *Bacillus* species (*B. subtilis* and *B. cereus*) [11, 12]. By the other side, the genus *Bacillus* has demonstrated the ability to produce phenolic acid decarboxylase that converts phenolic acids like p-coumaric acid and FA to some toxic compounds by the mechanism of resistance to aromatic compounds [40]. Finally, it should be mentioned that this is one of the first studies to quantify the capacity of bioconversion the FA to CAS by the agar plate.

5 Conclusions

The present study is the first report that describes quantitative methods that evidence the pectinolytic activity and the bioconversion of FA to aromatic compounds in *B. subtilis* IFVB and *B. cereus* IFVB isolated from *V. planifolia* Jacks ex Andrews during the Mexican vanilla bean curing process. The best isolated evaluated in this study was *B. subtilis* strain (7) from the Papantla region, due to it produced the highest PAI in stage 1 tested by the cup-plate using enzymatic extract. And the best isolated evaluated in the bioconversion test of FA to CAS was *B. cereus* strain (9) from the Papantla region, which produces the highest BFAI in stage 1. The results of the present study emphasize the role of both *Bacillus* species in the contribution of the aromatic profile from Mexican vanilla during the vanilla bean curing process. It will be useful in future studies to prove the development of aromatic compounds using *Bacillus* strains in mixed cultures to research their influence on FA assimilation.

Acknowledgements We appreciate the help we provided in obtaining vanilla beans from producers of *Vanilla planifolia* in Papantla Veracruz-Mexico in the collection of vanilla beans. Also, thanks to the Colegio de Ingenería de Alimentos, Facultad de Ingeniería Química and the Centro de Investigación en Ciencias Microbiológicas del Instituto de Ciencias, BUAP (CICM-ICUAP) for donating the funds for this research. During the time that this work was being completed, the author was awarded a Conacyt doctoral fellowship (Scholarship No. 40696).

Conflict of Interest The authors declare no conflict of interest.

References

1. Röling WF, Kerler J, Braster M, Apriyantono A, Stam H, van Verseveld HW (2001) Microorganisms with a taste for vanilla: microbial ecology of traditional Indonesian vanilla curing. Appl Environ Microbiol 67:1995–2003. https://doi.org/10.1128/AEM.67.5.1995-2003.2001
2. Havkin-Frenkel D, Belanger FC (2010) Handbook of vanilla science and technology. Wiley, pp 62, 82. https://doi.org/10.1002/9781119377320
3. Chen P, Yan L, Wu Z, Li S, Bai Z, Yan X, Wang N, Liang N, Li H (2016) A microbial transformation using *Bacillus subtilis* B7-S to produce natural vanillin from ferulic acid. Sci Rep 6:1–10. https://doi.org/10.1038/srep20400

4. Chen Y, Gu F, Li J, He S, Xu F, Fang Y (2015) Involvement of colonizing *Bacillus* isolates in glucovanillin hydrolysis during the curing of *Vanilla planifolia* Andrews. Appl Environ Microbiol 81:4947–4954. https://doi.org/10.1128/AEM.00458-15
5. Luna-Guevara JJ, Ruiz-Espinosa H, Herrera-Cabrera EB, Navarro-Ocaña A, Delgado-Alvarado A, Luna-Guevara ML (2016) Variedad de microflora presente en vainilla (*Vanilla planifolia* Jacks. ex Andrews) relacionados con procesos de beneficiado. Agroproductividad 9:3–9
6. Zhao Q, Wang H, Zhu Z, Song Y, Yu H (2015) Effects of *Bacillus cereus* F-6 on promoting vanilla (*Vanilla planifolia* Andrews.) plant growth and controlling stem and root rot disease. Agric Sci 6:1068–1078. https://doi.org/10.4236/as.2015.69102
7. Miller GL (1959) Modified DNS method for reducing sugars. Anal Chem 31:426–428
8. Mercimek Takcı HA, Turkmen FU (2016) Extracellular pectinase production and purification from a newly isolated *Bacillus subtilis* strain. Int J Food Prop 19:2443–2450. https://doi.org/10.1080/10942912.2015.1123270
9. Varghese L, Rizvi A, Gupta A (2013) Isolation, screening and biochemical characterization of pectinolytic microorganism from soil sample of Raipur city. J Biol Chem Res 30:636–643
10. Anjum F, Zohra RR, Ahmad M, Zohra RR (2019) Pectinase producers from rotten fruits and vegetable samples: isolation, screening and characterization. Int J Sci 8:54–59. https://doi.org/10.18483/ijSci.2033
11. Ito N, Itakura M, Eda S, Saeki K, Oomori H, Yokoyama T et al (2006) Global gene expression in *Bradyrhizobium japonicum* cultured with vanillin, vanillate, 4-hydroxybenzoate and protocatechuate. Microbes Environ 21:240–50. https://doi.org/10.1264/jsme2.21.240
12. Pundir S, Martin MJ, O'Donovan C (2017) UniProt protein knowledgebase. In: Wu C, Arighi C, Ross K (eds) Protein bioinformatics. Methods in molecular biology, vol 1558. Humana Press, New York, NY. https://doi.org/10.1007/978-1-4939-6783-4_2
13. Labuda I, Daphna HF, Faith CB (2010) Biotechnology of vanillin: vanillin from microbial sources. In: Handbook of vanilla science and technology, pp 299–331
14. Dunlap CA, Kwon SW, Rooney AP, Kim SJ (2015) *Bacillus paralicheniformis* sp. nov., isolated from fermented soybean paste. Int J Syst Evol Microbiol 65:3487–3492. https://doi.org/10.1099/ijsem.0.000441
15. Slepecky RA, Hemphill HE (2006) The genus *Bacillus*—nonmedical. In: The prokaryotes: volume 4: bacteria: *firmicutes, cyanobacteria*, pp 530–562
16. Tallent SM, Kotewicz KM, Strain EA, Bennett RW (2012) Efficient isolation and identification of *Bacillus cereus* group. J AOAC Int 95:446–451. https://doi.org/10.5740/jaoacint.11-251
17. Sambrook J, Russell DW (2006) The condensed protocols from molecular cloning: a laboratory manual (No. Sirsi) i9780879697723
18. Hasegawa H, Chatterjee A, Cui Y, Chatterjee AK (2005) Elevated temperature enhances virulence of *Erwinia carotovora* subsp. *carotovora* strain EC153 to plants and stimulates production of the *Quorum Sensing* signal, N-acyl homoserine lactone, and extracellular proteins. Appl Environ Microbiol 71:4655–4663. https://doi.org/10.1128/AEM.71.8.4655-4663.2005
19. Zamzuri NA, Abd-Aziz S, Rahim RA, Phang LY, Alitheen NB, Maeda T (2014) A rapid colorimetric screening method for vanillic acid and vanillin-producing bacterial strains. J Appl Microbiol 116:903–910. https://doi.org/10.1111/jam.12410
20. Mohandas A, Raveendran S, Parameswaran B, Abraham A, Athira RS, Kuruvilla Mathew A, Pandey A (2018) Production of pectinase from *Bacillus sonorensis* MPTD1. Food Technol Biotechnol 56:110–116. https://doi.org/10.17113/ftb.56.01.18.5477
21. Torimiro N, Adediwura VA, Ojo ST, Oluwadare AO, Okonji RE (2018) Pectinolytic activities of pectinase produced by some bacterial isolates cultured from deteriorating fruits. Niger J Biotechnol 35:91–98. https://doi.org/10.4314/njb.v35i2.12
22. Tumane PM, Tambe KS, Wasnik DD, Kolte NA (2018) Production of pectinase enzyme by pectinolytic bacteria isolated from fruit waste dumping soil samples. Int J Res Anal Rev 5(3):826–835. ISSN 2348-1269, Print ISSN 2349-5138
23. Drissi Kaitouni LB, Anissi J, Sendide K, El Hassouni M (2020) Diversity of hydrolase-producing halophilic bacteria and evaluation of their enzymatic activities in submerged cultures. Ann Microbiol 70:1–15. https://doi.org/10.1186/s13213-020-01570-z

24. Gophanea SR, Khobragadea CN, Jayebhayea SG (2021) Extracellular pectinase activity from *Bacillus cereus* GC subgroup A: isolation, production, optimization and partial characterisation. J Microbiol Biotechnol Food Sci 767–772. https://doi.org/10.15414/jmbfs.2016.6.2.767-772
25. Gu F, Chen Y, Fang Y, Wu G, Tan L (2015) Contribution of *Bacillus* isolates to the flavor profiles of vanilla beans assessed through aroma analysis and chemometrics. Molecules 20:18422–18436. https://doi.org/10.3390/molecules201018422
26. Hugouvieux-Cotte-Pattat N, Condemine G, Shevchik VE (2014) Bacterial pectate lyases, structural and functional diversity. Environ Microbiol Rep 6:427–440. https://doi.org/10.1111/1758-2229.12166
27. Pickaver AH (1977) Diagnostic agar plate techniques for testing pectinase-producing bacteria can give false negative results. FEMS Microbiol Lett 2:105–107. https://doi.org/10.1111/j.1574-6968.1977.tb00918.x
28. Soares MMCN, Da Silva R, Carmona EC, Gomes E (2001) Pectinolytic enzyme production by *Bacillus* species and their potential application on juice extraction. World J Microbiol Biotechnol 17:79–82. https://doi.org/10.1023/A:1016667930174
29. Kaur SJ, Gupta VK (2017) Production of pectinolytic enzymes pectinase and pectin lyase by *Bacillus subtilis* SAV-21 in solid state fermentation. Ann Microbiol 67:333–342. https://doi.org/10.1007/s13213-017-1264-4
30. Seo WT, Lim WJ, Kim EJ, Yun HD, Lee YH, Cho KM (2010) Endophytic bacterial diversity in the young radish and their antimicrobial activity against pathogens. J Korean Soc Appl Biol Chem 53:493–503. https://doi.org/10.3839/jksabc.2010.075
31. Graf N, Wenzel M, Altenbuchner J (2016) Identification and characterization of the vanillin dehydrogenase YfmT in *Bacillus subtilis* 3NA. Appl Microbiol Biotechnol 100:3511–3521. https://doi.org/10.1007/s00253-015-7197-6
32. Oumer OJ, Abate D (2018) Screening and molecular identification of pectinase producing microbes from coffee pulp. Biomed Res Int 1–8. https://doi.org/10.1155/2018/2961767
33. Singh A, Kaur A, Dua A, Mahajan R (2015) An efficient and improved methodology for the screening of industrially valuable xylano-pectino-cellulolytic microbes. Enzyme Res 1–8. https://doi.org/10.1155/2015/725281
34. Kavuthodi B, Thomas SK, Sebastian D (2015) Co-production of pectinase and biosurfactant by the newly isolated strain *Bacillus subtilis* BKDS1. Microbiol Res J Int 10:1–12. https://doi.org/10.9734/BMRJ/2015/19627
35. Khan F, Latif Z (2016) Molecular characterization of polygalacturonase producing bacterial strains collected from different sources. J Anim Plant Sci 26:612–618
36. Husseiny SM, Bayoumi RA, El-Gamal MS, Ahmad AI, Khashaba HM (2008) Production and purification of pectinase, avicelase and carboxymethyl cellulase by fermentation of agro-industrial wastes using *B. firmus* and *B. laterosporus*. N Egypt J Microbiol 19:326–352
37. Gallage NJ, Møller BL (2015) Vanillin–bioconversion and bioengineering of the most popular plant flavor and its de novo biosynthesis in the vanilla orchid. Mol Plant 8:40–57. https://doi.org/10.1016/j.molp.2014.11.008
38. Mishra S, Sachan A, Vidyarthi AS, Sachan SG (2014) Microbial production of 4-vinylguaiacol from ferulic acid by *Bacillus cereus* SAS-3006. Biocatal Biotransform 32:259–266. https://doi.org/10.3109/10242422.2014.974573
39. Mishra S, Panjiar N, Sachan A, Vidyarthi AS, Sachan SG (2017) Ferulic acid decarboxylase from *Bacillus cereus* SAS-3006: purification and properties. In: Applications of biotechnology for sustainable development. Springer, Singapore, pp 169–179
40. Park SC, Kwak YM, Song WS, Hong M, Yoon SI (2017) Structural basis of effector and operator recognition by the phenolic acid-responsive transcriptional regulator PadR. Nucleic Acids Res 45:13080–13093. https://doi.org/10.1093/nar/gkx1055

Evaluation of Biocompatibility of a Standardized Extract of *Agave angustifolia* Haw in Human Dermal Fibroblasts

Herminia López-Salazar, Jesús Santa-Olalla Tapia, Brenda Hildeliza Camacho-Díaz, Martha L. Arenas Ocampo, and Antonio R. Jiménez-Aparicio

Abstract *Agave angustifolia* Haw has been used as an ethnobotanical plant, for the treatment of various diseases. In addition, different biological activities attributed to its active compounds have been reported, but their biocompatibility has been little explored. The objective of this research work was to determine the in vitro cytotoxicity of the ethanolic extract of *A. angustifolia*. Methodology: Six different concentrations (0.01, 0.1, 1, 10, 100, 500 μg/ml) of the ethanolic extract of *A. angustifolia* were tested for 48 h of contact with the human dermal fibroblasts (HDF) to evaluate its cytotoxicity using the sulforhodamine (SRB) colorimetric test. On the contrary, HDF with concentrations of 100 and 500 μg/ml showed a high percentage of viability over the control group, more evident in the concentration of 500 μg/ml, which had a 24% increase in the fibroblast population. Conclusions: The ethanolic extract from *A. angustifolia* obtained by MAE, there was no evidence of cytotoxicity. It was able to be considered as a biocompatible material. Rather, it was shown to promote fibroblast proliferation.

Keywords Microwave-assisted extraction · β-sitosterol β-D-glucoside · Cytotoxicity · Human fibroblasts

1 Introduction

Medicinal plant species are being used in traditional medicine, including beneficial bioactive compounds with the primary goal of curing diseases in humans and animals. Ethopharmacology's goal is drug development for patient treatments and, by last, the traditional use of medicinal plants validation [1].

H. López-Salazar (✉) · B. H. Camacho-Díaz · M. L. A. Ocampo · A. R. Jiménez-Aparicio (✉)
Centro de Desarrollo de Productos Bióticos, Instituto Politécnico Nacional, P.O. Box 24, 62730 Yautepec, Morelos, Mexico

J. S.-O. Tapia
Facultad de Medicina, Universidad Autónoma del Estado de Morelos (UAEM), Calle Leñeros esquina Iztaccíhuatl s/n Col. Volcanes, 62350 Cuernavaca, Morelos, Mexico

K. L. Flores Rodríguez et al. (eds.), *Recent Trends in Sustainable Engineering*, Lecture Notes in Networks and Systems 297,
https://doi.org/10.1007/978-3-030-82064-0_9

The genus Agave is an example of medicinal plants, it is used to reduce or alleviate certain discomforts suffered by humans such as "heaviness of the stomach", "joint pain", "tooth pain", "bumps", "wounds healing" and "toxic animals stings or bites" [2]. Furthermore, the genus Agave is an important source of bioactive compounds, which can vary taking into account the type of Agave, age, environmental conditions and the extraction method used [3].

In addition, the literature has reported active compounds from different Agave species, such as fatty acids [4], fructans [5], flavonoids [6], homoisoflavonoids [4], phenols [7], volatile coumarins, tannins [8], long-chain alkanes [4], saponins, steroidal sapogenins [9] and sterols [10].

Furthermore, different pharmacological activities for example immunomodulatory, anti-inflammatory, antiparasitic and cytotoxic have been attributed to this genus [11–14]. However, biocompatibility tests of extracts of *A. angustifolia* have not currently been performed in human dermal fibroblasts.

For medical devices that are exposed in direct contact with human skin, biocompatibility testing is vitally important. These tests are established in ISO 10993 Biological Evaluation of Medical Devices, in which it mentions biological tests to evaluate the safety of biomaterials [15].

In order to detect in time some unwanted effect of the new biological medical devices, which are destined to be in contact with the human body, cytotoxicity tests are used. This is an in vitro test to examine whether the medical device, for example an extract obtained from a plant, could cause cell death due to a toxic leakage or through direct contact. The test protocols are set out in ISO 10993-5. Once it has been determined in vitro that the material will not cause cell death, the following test will be performed in an in vivo model to detect any skin irritation. Therefore, in vitro models are used first to examine the experimental product, before in vivo testing, especially when it is planned to be used for topical administration [16].

Likewise, different cell lines are frequently used in this type of assay, such as murine cell lines, for example, L929, Balb/c 3T3, and C3H-L, among others. In addition, murine cells from certain organs, such as the liver or spleen, can also be used. On the other hand, tests are also carried out with cells of human origin, among which are: HeLa and HaCaT cells; T lymphocytes, macrophages, keratinocytes, and fibroblasts from different places such as skin, lung, buccal mucosa, and the periodontal membrane. All the cells mentioned above can be used in the different biocompatibility tests, but it is recommended to use cells that are close to the human tissue or organ of interest. Therefore, human epidermal keratinocytes and fibroblasts would be the ideal cells to perform cytotoxicity and tolerance evaluations of biomaterials for topical administration on human skin. This is because they are cells that are actively involved in the wound healing process, inflammation, and the immune system [17].

Recently, there is a great variety of biomaterials obtained from medicinal plants, which is why the need arises to evaluate their safety, to implement them in the improvement of human health. In this case, the biocompatibility of biomaterials is an important component to determine their possible use in biomedicine [18].

Under these circumstances, the objective of this research was to analyze the cytotoxic activity of *A. angustifolia* extract on human dermal fibroblasts (HDF) in a contact period of 48 h by means of the SRB (sulforhodamine) assay, with the aim of helping the development of safe topical drugs treatments.

2 Materials and Methods

2.1 *Plant Material*

The part of the plant used for this work was the head (stem) of 5-year-old *A. angustifolia*, which has a registered biological identity. It was acquired in Yautepec, Morelos, Mexico, located within the following geographic coordinates (18° 49′ 33.3″ N and 99° 06′ 21.98″ W; 1120 m above mean sea level) in April 2019.

2.2 *Extraction Technique*

32.4 kg of fresh material sectioned from the head of *A. angustifolia* were obtained. Subsequently, the plant material was dried in an oven (40 °C) for 48 h. Then, the plant material was ground in a mill (M-200, INMIMEX, Mexico) obtaining a plant material size of 0.074 mm (200 mesh sieve). The final yield obtained was 3965 kg of dry plant material.

The microwave extraction method (MAE) was used to obtain the ethanolic extract (EE) of *A. angustifolia* (extraction time of 5 s), this procedure was described by López-Salazar in 2019 [10]. The EE was stored in a glass vial in the dark at −18 °C until tested.

2.3 *Cell Culture*

Human Dermal fibroblasts (Primary Dermal Fibroblast; Normal, Human, Adult (HDFa) (ATCC® PCS-201-010™)) were used. First, the cells contained in the cryopreservation tubes were thawed, counted and seeded in Petri dishes for cell culture of 100 mm in diameter at a density of 1×10^6 cells/cm^2, in an atmosphere of 5% CO_2, at 37 °C, in DMEM culture medium with antibiotic, antifungal and 10% fetal bovine serum. When the fibroblasts reached 80% of confluence were propagated until 80%, they were separated using 0.25% trypsin. Later, to identify and count living and dead cells, in the Neubauer chamber, trypan blue was used as a vital stain. Counting was carried out by averaging the values obtained from 5 quadrants of the Neubauer chamber. After counting the quadrants, the number of living cells was determined by the following equation:

$$Total\,living\,cells = \frac{(Summation\,living\,cells)(20{,}000)(Volume\,of\,media\,used)}{5} \quad (1)$$

Subsequently, the number of cells needed to carry out the cytotoxicity assay was calculated in a 96-well multi-well box.

2.4 In Vitro Cell Biocompatibility

For this test, the number of cells used was 500.00 cells per well per culture box of 100 mm diameter for their propagation, which allows reaching 80% confluence after approximately 4 days, for the test. First, a 96-well plate was used, in DMEM culture medium with the following antibiotics (100 µg/µL of streptomycin, 100 U/µL of penicillin), antifungal (0.25 µg/mL of fungizone) and enriched with bovine serum 10% fetal, then incubated at 37 °C in a 5% CO_2 atmosphere. For the beginning of the evaluation of the cytotoxicity of the EE of *A. angustifolia*, different dilutions of the EE were applied per well (0.01, 0.1, 1, 10, 100, 500 µg/ml). At 48 h the EE was eliminated with the culture medium from each well. Subsequently, the cultures were fixed by applying in each well of 100 µl with 10% trichloroacetic acid (TCA), this for 1 h at 4 °C, followed by 5 washes with distilled water. Drying was continued for 12 h at room temperature. The cytotoxicity test was carried out by the SRB colorimetric method; therefore, the cultures were stained with SRB for 30 min at room temperature. After each well was washed with 1% acetic acid (200 µl), then the plates were dried by 30 min in a flow hood. The SRB dye was then removed from each plate by adding 50 µl per well of 1% TRIS stock solution and brought to a SINERGY2 ELISA plate reader using Gene5 Data Analysis Subcultures software, to measure absorbance (wavelength 570 nm). The absorbances obtained from each treatment were compared with the fibroblast culture without the application of the extract (control group).

The viability was determined with the following equation:

$$\%\,Viability = \frac{100 \times OD_{570e}}{OD_{570b}} \quad (2)$$

OD570e = average of the optical density measurement of the concentrations of the extracts at 100% of the test sample.

OD570b = average of the optical density measurement of the concentrations of the control group.

It is equally important to mention that, when there is a decrease in the number of proteins in the sample of interest, it is correlated with a decrease in the number of living cells. This decrease is related to the optical density detected at 570 (amount of SRB formed).

2.5 *Statistical Analysis*

The statistical data obtained were analyzed using a variance factor (ANOVA). The Holm-Sidak method was applied to determine multiple comparisons versus control group. Significant differences were considered when ($p < 0.001$). Statistical analysis of the results was performed using SigmaPlot 11.0 software (Systat Software Inc., San José, CA, USA) [19].

3 Results and Discussion

3.1 *Cellular Biocompatibility*

The results obtained after the treatment period of 48 h, showed that the different concentrations (0.01, 0.1, 1.0, 10, 100, 500 μg/ml) of the ethanolic extract of *A. angustifolia* obtained by microwaves, in an extraction time of 5 s, did not show any cytotoxic effect in human dermal fibroblast cultures. It is important to mention that the SRB staining method is frequently used as a screening method to measure cytotoxicity in vitro. For this reason, this staining method was used to evaluate the cytotoxic capacity of *A. angustifolia* EE on HDF [20]. Optical density units are an indicator of cytotoxic activity, to put it another way, the higher the optical density, the higher the living cells, this is because the SRB dye adheres to the proteins of living cells. The results obtained are shown in Fig. 1, which shows that the optical densities obtained in this experimental work are above the control group used.

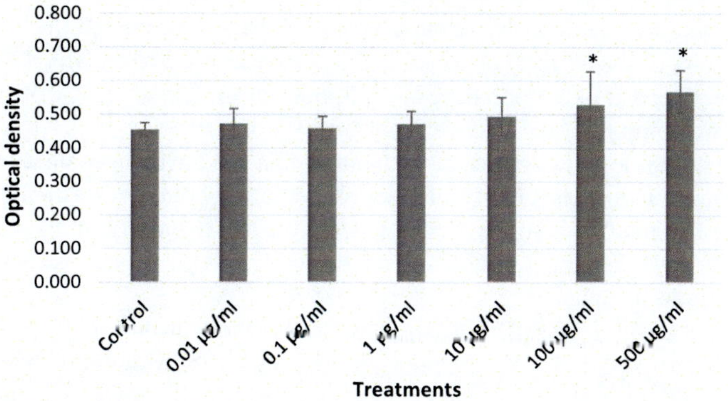

Fig. 1 Cytotoxic evaluation of the ethanolic extract of *A. angustifolia* in HDF after 48 h of incubation. n = 18. Statistical analysis showed that there is a significant difference between the control and treatments. $^*p \leq 0.001$. Using the Holm-Sidak method

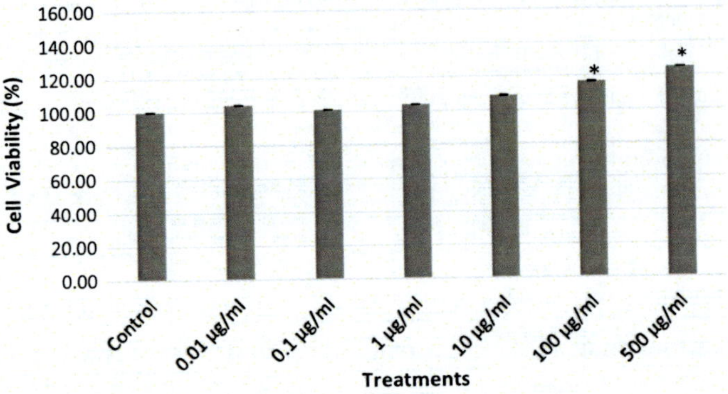

Fig. 2 Viability of HDF treated with ethanolic extract of *A. angustifolia* after 48 h of incubation. n = 18. Statistical analysis showed that there is a significant difference between the control and treatments. **p* ≤ 0.001. Using the Holm-Sidak method

This means that an increase in the HDF population was observed with the application of the different concentrations of the EE of *A. angustifolia*, which may suggest an increase in the proliferation or survival of the fibroblasts.

On the other hand, the effect obtained from the 6 different applied concentrations of the ethanolic extract of *A. angustifolia* obtained by MAE did not affect the cell viability of HDF. But on the contrary, as can be seen in Fig. 2, the fibroblast viability percentage is above the control in the highest doses (100 and 500 μg/ml). Furthermore, this increase in the percentage of viability becomes more noticeable at the concentration of 500 μg/ml, in which a 24% increase in the cell population was observed.

Although previous results obtained in our research group suggested that the standardized extract based on β-sitosterol β-D-glucoside (BSSG) (124.76 mg/g dry weight of the extract) obtained in 5 s of extraction time by MAE, has certain advantages such as extraction time and it is an environmentally friendly extraction method, if we compare it with conventional extraction methods [10].

Likewise, it is important to mention that the advantages of using a standardized extract such as the quality, efficacy and reliability of phytomedicine and this is due to the fact that the active ingredients are more stable and sometimes the yield is higher [21]. One advantage of this research work was to use an extract based on BSSG (standardized).

In addition, the BSSG has a research report on fibroblast proliferation, which showed potential wound healing properties by showing an increase in the viability of L929 fibroblasts and their migration, also stimulating the production of collagen, a rhizome of *Boesenbergia kingii* extract [22].

In like manner, there are research articles on fibroblast proliferation from natural products. Examples of them we can name the work of Mazumdar in 2021 [23], which showed the 32% ethanolic extract obtained by the maceration extraction method of *Annona reticulata* L. stimulated the proliferation of dermal fibroblasts, at 48 h, with a concentration of 20 μg/ml (127,023 ± 5.1%) using the MTT assay. Furthermore, it showed a significant migration of fibroblasts and keratinocytes.

Another research work, which is important to mention, is that carried out by Haddadi in 2019 [24], this study was carried out to evaluate the biological activity of the hydroalcoholic extract of *Scrophularia striata* on the viability, proliferation, and migration of endothelial cells of the umbilical vein and in human dermal fibroblasts.

They evaluated seven different concentrations (10, 50, 100, 200, 400, 800 and 1000 μg/ml) of the ethanolic extract. They used the MTT test to investigate cell proliferation and viability. For human dermal fibroblast cell lines, they used three times of exposure to the extract (24, 48 and 72 h). The hydroalcoholic extract of *S. striata* did not show toxicity in human dermal fibroblasts at a concentration below 800 μg/ml, however, their viability decreased significantly compared to the control group, when exposed to 800 μg/ml after 48 and 72 h.

Furthermore, the 1000 μg/ml concentration of hydroalcoholic extract was shown to be toxic to human dermal fibroblast cells. With these results they demonstrated that the hydroalcoholic extract of *S. striata* has both proliferative and migratory activity on human skin fibroblasts, at concentrations below 800 μg/ml.

Among the advantages of conducting in vitro tests is the reliable and reproducible detection of cell death, as well as the timely detection of negative effects on cell functions. Being standardized studies, they are considered a reproducible method, so they are suitable tests to evaluate the response of different cells to some biomaterials. The different methodologies for cytotoxicity assays are detailed in various standards, one of which is the International Organization for Standardization (ISO) 10993-5: 2009.3,4T [25].

Likewise, considering the cytotoxicity test by ISO: 10993-5: 2009.3,4T [25], which establishes that a product has cytotoxic potential when the % of the viability of the cell culture is <70%, as long as it is compared to the control group.

If we consider those established in the ISO: 10993-5: 2009.3,4T, it is confirmed that all concentrations used in this work of the ethanolic extract of *A. angustifolia* are not cytotoxic (Table 1).

Table 1 Cytotoxic evaluations of ethanolic extract of *A. angustifolia* in HDF after 48 h of incubation

Extract concentrations	% Viability	Cytotoxicity
Control	100.00 ± 3.96	Viable
0.01 μg/ml	103.84 ± 8.67	Viable
0.1 μg/ml	100.89 ± 7.04	Viable
1 μg/ml	103.52 ± 7.43	Viable
10 μg/ml	108.52 ± 10.53	Viable
100 μg/ml	116.36 ± 17.18*	Viable
500 μg/ml	124.87 ± 10.40*	Viable

*Multiple comparisons against the control group. Holm-Sidak method

$n = 18, p \leq 0.001$

After comparing our data with the scientific literature, we investigated the biocompatibility of the ethanolic extract of *A. angustifolia* obtained by MAE, and found that it was not cytotoxic, instead, there was an increase in the cell population of HDF.

This result of increased cell viability in human dermal fibroblasts is important because these cells are the main components of the dermis. In addition, they have an important function in the structural support of tissues, they also collaborate in the performance of several fundamental activities for humans, such as wound healing, and participate in the regulation of inflammation and proliferation [26].

Likewise, a wound is an injury that damages the continuity of the skin or other tissues, which can cause structural and functional damage [27]. On the other hand, the healing of a wound is healing, which is a continuous and dynamic process, which includes 4 phases, which have the characteristic that they overlap. These phases include hemostasis, inflammation, proliferation, and remodeling [28]. But there are different factors that cause the stagnation of one or more phases, leading to the development of a chronic wound, which is difficult to heal. These types of injuries occur frequently in people with diabetes or vascular disease [29].

If they do not have the proper treatment, they can be complicated up to an amputation. In the continuous search for new treatments, different research works have shown beneficial effects, among these are studies of plants with healing effects [30].

Therefore, the result obtained in this research provides great knowledge for the management of chronic wounds, considering the standardized ethanolic extract of *A. angustifolia* as a potential agent to facilitate the healing of chronic wounds, through the proliferation of human dermal fibroblasts.

4 Conclusions

According to the evidence shown in this work, the standardized ethanolic extract of *A. angustifolia* obtained by MAE, there was no evidence of cytotoxicity, so it could be considered as a biocompatible material. Likewise, it can be argued that this extract can be considered as a candidate to accelerate the wound healing process due to the increase in fibroblasts shown. Other related evidence needs to be explored to promote wound healing, such as migration fibroblast assays in vitro, keratinocyte assays, and in vivo models, among others.

Acknowledgements This work was funded by project SIP 20196656 of the Instituto Politécnico Nacional and through a basic grant from the Consejo Nacional de Ciencia y Tecnología (CONACyT), Mexico for the Ph.D. studies of Herminia López Salazar, in the CEPROBI, Instituto Politécnico Nacional. As well as, the BEIFI scholarship was provided by the Instituto Politécnico Nacional, Mexico.

References

1. Anand U, Jacobo-Herrera N, Altemimi A, Lakhssassi N (2019) A comprehensive review on medicinal plants as antimicrobial therapeutics: potential avenues of biocompatible drug discovery. Metabolites 9(11):258. https://doi.org/10.3390/metabo9110258
2. Flores NB, Araiza PLS (2012) El mezcal en Sonora, México, más que una bebida espirituosa. Etnobotánica de *Agave angustifolia* Haw. Estud Soc Rev aliment contemp desarro reg 2:173–197
3. López-Salazar H, Martínez-Velarde R, Arenas-Ocampo ML, Camacho-Díaz B, Jiménez-Aparicio AR (2019) Identification and quantification of phytosterols of an ethanolic extract obtained by microwave-assisted extraction. In: Integral and sustainable use of Agave, pp 113–117
4. Morales-Serna JA, Jiménez A, Estrada-Reyes R, Marquez C, Cárdenas J, Salmón M (2010) Homoisoflavanones from *Agave tequilana* Weber. Molecules 15(5):3295–3301. https://doi.org/10.3390/molecules15053295
5. Velázquez-Martínez JR, González-Cervantes RM, Hernández-Gallegos MA, Mendiola RC, Aparicio ARJ, Ocampo MLA (2014) Prebiotic potential of *Agave angustifolia* Haw fructans with different degrees of polymerization. Molecules 19(8):12660–12675. https://doi.org/10.3390/molecules190812660
6. Barriada-Bernal LG, Almaraz-Abarca N, Delgado-Alvarado EA, Gallardo-Velázquez T, Ávila-Reyes JA, Torres-Morán MI, Herrera-Arrieta Y (2014) Flavonoid composition and antioxidant capacity of the edible flowers of *Agave durangensis* (Agavaceae). CYTA J Food 12(2):105–114. https://doi.org/10.1080/19476337.2013.801037
7. Almaraz-Abarca N, Delgado-Alvarado EA, Ávila-Reyes JA, Uribe-Soto JN, González-Valdez LS (2013) The phenols of the genus Agave (Agavaceae). J Biomater Nanobiotechnol 4(3a):9–16. https://doi.org/10.4236/jbnb.2013.43A002
8. Ahumada-Santos YP, Montes-Avila J, De Jesús U-B, Díaz-Camacho SP, López-Angulo G, Vega-Aviña R, Delgado-Vargas F (2013) Chemical characterization, antioxidant and antibacterial activities of six Agave species from Sinaloa, Mexico. Ind Crops Prod 49:143–149. https://doi.org/10.1016/j.indcrop.2013.04.050
9. Gutiérrez-Nava ZJ, Jiménez-Aparicio AR, Herrera-Ruiz ML, Jiménez-Ferrer E (2017) Immunomodulatory effect of *Agave tequilana* evaluated on an autoimmunity like-SLE model induced in Balb/c mice with pristane. Molecules 22(6):848. https://doi.org/10.3390/molecules22060848
10. López-Salazar H, Camacho-Díaz BH, Ávila-Reyes SV, Pérez-García MD, González-Cortazar M, Arenas Ocampo ML, Jiménez-Aparicio AR (2019) Identification and quantification of β-Sitosterol β-d-glucoside of an ethanolic extract obtained by microwave-assisted extraction from *Agave angustifolia* Haw. Molecules 24(21):3926. https://doi.org/10.3390/molecules24213926
11. Hernández-Valle E, Herrera-Ruiz M, Salgado G, Zamilpa A, Ocampo MLA, Aparicio ARJ, Jiménez-Ferrer E (2014) Anti-inflammatory effect of 3-O-[(6'-O-palmitoyl)-β-D-glucopyranosyl sitosterol] from *Agave angustifolia* on ear edema in mice. Molecules 19(10):15624–15637. https://doi.org/10.3390/molecules191015624
12. Monterrosas-Brisson N, Ocampo MLA, Jiménez-Ferrer E, Jiménez-Aparicio AR, Zamilpa A, Gonzalez-Cortazar M, Herrera-Ruiz M (2013) Anti-inflammatory activity of different Agave plants and the compound Cantalasaponin-1. Molecules 18(7):8136–8146. https://doi.org/10.3390/molecules18078136
13. Botura MB, Silva GD, Lima HG, Oliveira JVA, Souza TS, Santos JDG, Batatinha MJM (2011) In vivo anthelmintic activity of an aqueous extract from sisal waste (*Agave sisalana* Perr.) against gastrointestinal nematodes in goats. Vet Parasitol 177(1–2):104–110. https://doi.org/10.1016/j.vetpar.2010.11.039
14. Mina SA, Melek FR, Abdel-Khalik SM, Gabr NM (2013) Two steroidal saponins from *Agave franzosinii* and *Agave angustifolia* leaves and biological activities of *Agave franzosinii*. J Nat Prod 6:188–197

15. Bruinink A, Luginbuehl R (2011) Evaluation of biocompatibility using in vitro methods: interpretation and limitations. In: Tissue engineering III: cell-surface interactions for tissue culture. Springer, Berlin, Heidelberg, pp 117–152. https://doi.org/10.1007/10_2011_111
16. Ramakrishna S, Tian L, Wang C, Liao S, Teo WE (2015) Safety testing of a new medical device. In: Medical devices, pp 137–153
17. Wiegand C, Hipler UC (2009) Evaluation of biocompatibility and cytotoxicity using keratinocyte and fibroblast cultures. Skin Pharmacol Physiol 22(2):74–82. https://doi.org/10.1159/000178866
18. Konovalova MV, Markov PA, Durnev EA, Kurek DV, Popov SV, Varlamov VP (2017) Preparation and biocompatibility evaluation of pectin and chitosan cryogels for biomedical application. J Biomed Mater Res Part A 105(2):547–556. https://doi.org/10.1002/jbm.a.35936
19. SigmaPlot for Windows Version 11.0, Build 11.1.0.102 (2018) Systat Software, Inc. Wpcube GmbH, Germany
20. Şeker-Karatoprak G, İlgün S, Koşar M (2017) Phenolic composition, anti-inflammatory, antioxidant, and antimicrobial activities of *Alchemilla mollis* (Buser) Rothm. Chem Biodivers 14(9). https://doi.org/10.1002/cbdv.201700150
21. Patnala S, Kanfer I (2021) Quality control, extraction methods, and standardization: interface between traditional use and scientific investigation. In: Herbal medicine in andrology. Elsevier, pp 175–187
22. Sudsai T, Wattanapiromsakul C, Tewtrakul S (2016) Wound healing property of isolated compounds from *Boesenbergia kingii* Rhizomes. J Ethnopharmacol 184:42–48. https://doi.org/10.1016/j.jep.2016.03.001
23. Mazumdar S, Ghosh AK, Dinda M, Das AK, Das S, Jana K, Karmakar P (2019) Evaluation of wound healing activity of ethanol extract of *Annona reticulata* L. leaf both in vitro and in diabetic mice model. J Tradit Complement Med. https://doi.org/10.1016/j.jtcme.2019.12.001
24. Haddadi R, Tamri P, Jooni FJ (2019) In vitro wound healing activity of *Scrophularia striata* hydroalcoholic extract. S Afr J Bot 121:505–509. https://doi.org/10.1016/j.sajb.2019.01.002
25. ISO 10993-5:2009(E) International standard ISO 10993-5, third edition 2009-06-01, biological evaluation of medical devices—part 5: tests for in vitro cytotoxicity
26. Stunova A, Vistejnova L (2018) Dermal fibroblasts—a heterogeneous population with regulatory function in wound healing. Cytokine Growth Factor Rev 39:137–150. https://doi.org/10.1016/j.cytogfr.2018.01.003
27. Agyare C, Dwobeng AS, Agyepong N, Boakey YD, Mensah KB, Ayande PG, Adarkwa-Yiadom M (2013) Antimicrobial, antioxidant, and wound healing properties of *Kigelia africana* (Lam.) Beneth. and *Strophanthus hispidus* DC. Adv Pharmacol Sci 2013. https://doi.org/10.1155/2013/692613
28. Guo S, DiPietro LA (2010) Factors affecting wound healing. J Dent Res 89(3):219–229. https://doi.org/10.1177/0022034509359125
29. Han G, Ceilley R (2017) Chronic wound healing: a review of current management and treatments. Adv Ther 34(3):599–610. https://doi.org/10.1007/s12325-017-0478-y
30. Budovsky A, Yarmolinsky L, Ben-Shabat S (2015) Effect of medicinal plants on wound healing. Wound Rep Reg 23(2):171–183. https://doi.org/10.1111/wrr.12274

Stochastic Identification and Kalman Filter for Blood Glucose Estimation

Jesus R. Tavarez, Irma Y. Sanchez, Victor A. Maldonado, Martin Montes, and Raul A. Ortiz

Abstract Monitoring of blood glucose (BG) in diabetic patients is of utmost importance and an essential requirement in the development of an artificial pancreas for diabetes treatment based on the supply of exogenous insulin (EI). The algorithms of practical utility for an automated insulin infusion must consider discrete and low-order dynamic models for their efficient implementation and continuous execution. A virtual patient, based on a 19th-order nonlinear continuous physiological model, is used to analyze and represent the dynamics of glucose metabolism by stochastic identification. The resulting third-order linear discrete models strike an appropriate balance between accuracy and complexity. These models are used to program a discrete Kalman filter to predict BG in a virtual patient, taking advantage of its capabilities for handling measurement uncertainty and limitations in modeling accuracy. Tuning the parameters of the Kalman filter, by calculated process noise and average current BG meter accuracy, is discussed in this work and applied to evaluate its effect. Tuned noise parameters improve the performance of the Kalman filter during carbohydrate intake, EI delivery, random variation of measurement parameters, and measurement loss, achieving improvements of more than 50% with respect to the use of unquantified process and measurement uncertainty.

J. R. Tavarez · I. Y. Sanchez (✉) · V. A. Maldonado · M. Montes · R. A. Ortiz
Universidad Politecnica de Aguascalientes, Aguascalientes, Mexico
e-mail: irma.sanchez@upa.edu.mx; isanchez@ingmt.com

J. R. Tavarez
e-mail: rafael.tavarez@upa.edu.mx

V. A. Maldonado
e-mail: victor.maldonado@upa.edu.mx

M. Montes
e-mail: martin.montes@upa.edu.mx

R. A. Ortiz
e-mail: raul.ortiz@upa.edu.mx

I. Y. Sanchez
Ingenieria Mecatronica, S.A., Aguascalientes, Mexico

K. L. Flores Rodríguez et al. (eds.), *Recent Trends in Sustainable Engineering*,
Lecture Notes in Networks and Systems 297,
https://doi.org/10.1007/978-3-030-82064-0_10

Keywords Diabetes · Kalman filter · Stochastic identification · Virtual patient

1 Introduction

Diabetes Mellitus is a chronic disease characterized by high glucose levels, due to inadequate or no insulin secretion, or low insulin sensitivity of the body, factors enhanced by common nowadays sedentary lifestyle [1, 2]. According to statistics by the International Diabetes Federation, there were 463 million people, 20–79 years old, around the world with some type of diabetes in 2019, and this number is predicted to increase in 51% by the year 2045 [3].

Diabetes constitutes a serious health problem due to its immediate and long-term effects, and its social and economic repercussions, on family, national and global scales. Different strategies for the regulatory control of blood glucose (BG) have been proposed, however, the current challenges are the complexity of the dynamics of the physiological process of glucose-insulin and the uncertainty in the instrumentation, particularly in measurement. These conditions make the use of the Kalman filter suitable.

This research details the definition of a low-order linear dynamic model for the prediction phase, and for the optional calculation or tuning of the parameters required for the update phase of the discrete Kalman filter. The latter is contrasted against the use of noncomputed preestablished filter parameters. The model has one input for exogenous insulin (EI) supply rate and one output for BG, and is obtained by stochastic identification. A virtual patient is considered, based on the physiological model of Sorensen [4], which is highly detailed, widely used and consistent with other proposals for the in silico evaluation of diabetes technologies [5].

Within the context of diabetes, the Kalman filter has been used for different purposes associated or not with BG control schemes, and incorporating different options of types of dynamic models for its prediction phase.

The Kalman filter has been implemented for the prediction of blood insulin based on glucose measurements. Physiological models of minimal (Bergman model, for example) and intermediate complexity (such as the Hovorka model) are often used [6]. The prediction of blood insulin can be based on particular models of insulin pharmacokinetics, that is, restricted to one aspect of the complete modeling of glucose and insulin physiology, or based on approximate models of relatively high order for the explicit representation of blood insulin as a state of the system. The purpose of using models of the lowest possible order implies managing states that represent mixed physiological conditions.

The approximation of linear models of ninth order to the physiological process of glucose has been proposed, as a basis for developing control strategies using the Kalman filter [7]. The Kalman filter has been used as a fixed-gain state estimator and part of the design of a continuous time BG linear quadratic regulator system using a ninth-order state space model [8]. However, in addition to a discrete design, the order of the model can be further reduced for more efficient computational implementations, as discussed in this work, without resorting to a fixed Kalman gain that subtracts the ability to represent uncertainties and prevents the optimal filter performance.

The Kalman filter has also been used as part of a discrete bihormonal control strategy (model predictive control for insulin delivery and proportional derivative control for glucagon infusion) using ARMAX (autoregressive with moving average error and external input) linear models of second order containing a stochastic term for unmodeled effects [9]. In this case, the ARMAX model increases the complexity of the filter prediction model by including the error due to factors not considered, which are already represented in the process covariance matrix or process/dynamic noise. In this research, ARX (autoregressive with external input) models are appropriate under the premise that the Kalman filter uses a sufficient dynamic noise parameter.

Regarding applications related to BG sensors, the use of the extended Kalman filter is reported for nonlinear calibration, compensation of measurement errors and generation of alarms, as intelligent functions [10, 11]. In [12], the Kalman filter is based on the approximation of a third-order ARX model with predefined adjustment factors to adapt only the coefficient of a unique input term to a particular patient, and predetermined filter parameters for alerting of hypo or hyperglycemia, using a prediction horizon of 30 min and sample times of 5 and 15 min. In contrast to [12], in this research, the discrete linear models have a continuous counterpart, all coefficients are calculated based on the dynamics of the virtual patient, and the Kalman filter parameters are both predefined and calculated, for comparison. Similarly to [12], third-order models are selected in this work for the Kalman filter implementation. BG prediction may only use discrete identification based on recursive least squares for an adaptive modification of the parameters of an ARMAX model, and the detection of significant changes in these parameters to alert the patient [13]. The research reported in [14] performs an adaptive identification, where even the order of the system is determined online. Unlike [13, 14], the herein presented work does not perform adaptive identification, letting the fixed model limitation effect to be managed by the process noise parameter. Nonlinear or high gain observers have also been proposed as alternatives to the use of Kalman filters for the deterministic estimation of states [15].

This article is organized as follows: Sect. 2 specifies the software and methods used to model the virtual patient, identify the stochastic model of the process, design and test the Kalman filter; Sect. 3 presents the results obtained with different sampling times, disturbances and measurement conditions; finally, Sect. 4 concludes on the issues of implementation of the discrete filter and future work.

2 Resources and Methods

2.1 *Software Tools*

This work was carried out with MATLAB® R2019a software and Simulink®. In MATLAB®, the discrete identification and evaluation of the mean square error, *mse*, of the identified models are developed. Simulink® allows the integration of the continuous system that implements the virtual patient, and the discrete Kalman filter, allowing evaluation in very short real times.

2.2 *Physiological Model for Virtual Patient*

Sorensen's physiological model (see Fig. 1) consists of a set of 19 continuous differential equations of mass balances, to represent blood flow, and the metabolic process that causes the addition or removal of glucose, insulin and glucagon in different parts of the body. This model represents a 70 kg adult male. Compartments are defined for the liver, brain, heart, kidneys, peripheral system (bones and muscles), and digestive system [4]. First-order dynamics are added to the physiological model for the BG meter (with an average time constant of 0.25 min and unit gain), and for the absorption of glucose from food in the digestive compartment (time constant of 60 min and unit gain), raising the order of the virtual system to 21.

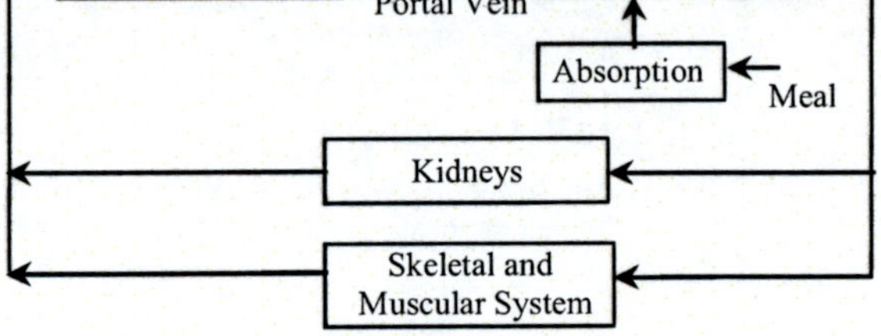

Fig. 1 Virtual patient represented by the Sorensen model with glucose measurement and gastro-intestinal glucose absorption dynamics

2.3 Stochastic Identification

The virtual patient allows the exploration of the metabolic response using any type of experiment not necessarily based on clinical tests. Although pulse signals and differential compensation signals may be used for approaching the practice of an oral glucose tolerance test [16] or a glucose clamp test [17], respectively, a richer variation of input signals is important for a proper stochastic identification.

A pseudo-random binary sequence (PRBS) type test [18] is adapted for discrete identification. The EI supply input is assigned as 0 or 15 mU/min creating pulses of different duration, according to the pseudorandom signal generated from a linear feedback shift binary register (Fig. 2). The XOR (exclusive OR) logic function is performed with two specific bits of the register, and the result is introduced to the left, displacing the bit values to the right, and producing the output of the bit at the right end of the register. If the register output is zero, the EI rate input remains at zero, but if it is one, the EI rate input value is made equal to 15 mU/min. The length, N, of the register determines the complete input signal sequence (a set of $2^N - 1$ values). Only one sequence is applied to limit the duration of the test. Two different sample times were used: for the sample time $T = 1$ min, $N = 8$ was chosen, and for $T = 15$ min, the register size was $N = 4$.

During the application of the PRBS test to the physiological model, EI supply rate and BG data pairs were obtained. The initial steady-state values were subtracted from the data. Then the set of deviation values were processed with the recursive least squares technique for the determination of the ARX model coefficients:

$$
\begin{aligned}
y(n) = a_1(n)y(n-1) + \cdots + a_{na}(n)y(n-na) + b_1(n)u(n-1-d) \\
+ \cdots + b_{nb}(n)u(n-nb-d)
\end{aligned}
\tag{1}
$$

where y is BG deviation value and u is EI supply rate.

The structure of the model must be specified, and is given by ARX(na, nb, d), where na is the number of autoregressive coefficients (a_i), nb is the number of external input coefficients (b_i), and d is the integer factor that multiplied by the sample time produces an approximation to the dead time of the process. The structures are limited to the cases where there is a continuous counterpart of the model, that is, with nb equal to na or to $na + 1$.

Although the identification was made offline, the recursive least square error algorithm [19] was used because of its efficient implementation. Equations (2–9) show the calculation procedure of the model coefficient vector $\theta(n) = [a_1, a_2, \ldots, a_{na}, b_1, b_2, \ldots, b_{nb}]'$, with fixed positive adjustment parameters δ and

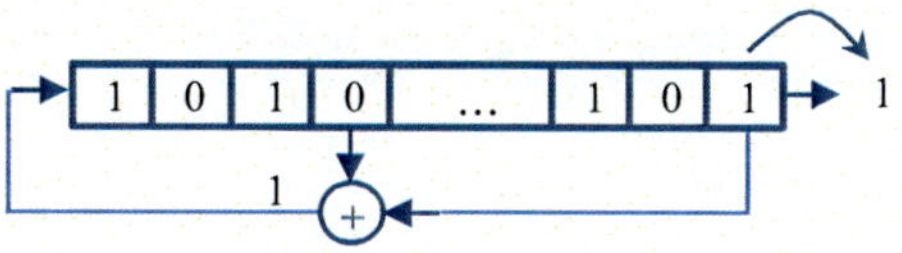

Fig. 2 Feedback and output of a linear shift register for producing a pseudorandom sequence

λ, model error ξ, and correction gain k_{LS}. For initialization or step 0,

$$\theta(0) = 0 \tag{2}$$

$$P(0) = \delta^{-1} I \tag{3}$$

Subsequent calculations in terms of the iteration index n (from $nb + d$ to $2^N - 1$) are

$$v(n) = [y(n-1), \ldots, y(n-na), u(n-1-d) \ldots u(n-nb-d)]' \tag{4}$$

$$Q(n) = P(n-1)v(n) \tag{5}$$

$$k_{LS}(n) = \frac{Q(n)}{\lambda + v'(n)Q(n)} \tag{6}$$

$$\xi(n) = y(n) - \theta'(n-1)v(n) \tag{7}$$

$$\theta(n) = \theta(n-1) + k_{LS}(n)\xi(n) \tag{8}$$

$$P(n) = \lambda^{-1}P(n-1) - \lambda^{-1}k_{LS}(n)v'(n)P(n-1) \tag{9}$$

Once the ARX model coefficients have been obtained, the discrete transfer function that relates BG with EI supply rate, in terms of $Y(z)$ and $U(z)$, respectively, can be defined by

$$\frac{Y_{aux}(z)}{U(z)} = \frac{d_1 z^{-1} + d_2 z^{-2} + \cdots + d_m z^{-m}}{1 - c_1 z^{-1} - c_2 z^{-2} - \cdots - c_m z^{-m}} \tag{10}$$

$$Y(z) = z^{-d} Y_{aux}(z) \tag{11}$$

where $m = nb$, a_i matches c_i, $c_i = 0$ for $i = na + 1, \ldots, nb$, and b_i matches d_i.

From the transfer function, the state space representation can be proposed as

$$x(k+1) = \begin{bmatrix} 0\ 1\ 0 & \cdots & 0 \\ 0\ 0\ 1 & \cdots & 0 \\ \vdots\ \vdots\ \vdots & \cdots & \vdots \\ c_m\ c_{m-1}\ c_{m-2} & \cdots & 1 \end{bmatrix} x(k) + \begin{bmatrix} 0 \\ 0 \\ \vdots \\ 1 \end{bmatrix} u(k) \tag{12}$$

$$y_{aux}(k) = \begin{bmatrix} d_m\ d_{m-1}\ d_{m-2} & \cdots & d_1 \end{bmatrix} x(k) \tag{13}$$

$$y(k) = y_{aux}(k - d) \tag{14}$$

where the state matrix and input vector in (12) are designated as A and B, respectively, and the output vector in (13) is represented by C.

2.4 Kalman Filter

The Kalman filter predicts the state of a process, $x(k)$, and corrects the prediction by considering the difference between a measurement, z(k), and the predicted output, $y(k)$ [19]. The filter consists of two groups of equations. The measurement update or correction equations are

$$K(k) = P(k, k-1)C^T\left[CP(k, k-1)C^T + R\right]^{-1} \tag{15}$$

$$\hat{x}(k, k) = \hat{x}(k, k-1) + K(k)\left(z_k - C\hat{x}(k, k-1)\right) \tag{16}$$

$$P(k, k) = (1 - K(k)C)P(k, k-1) \tag{17}$$

where K is the Kalman filter gain, P is the estimation covariance, R is the measurement uncertainty, the first index k refers to the present discrete time, and the second index $k - 1$ indicates the use of available data prior to k. The group of equations for time update or prediction is given by

$$\hat{x}(k+1, k) = A\hat{x}(k, k) + Bu(k) \tag{18}$$

$$P(k+1, k) = AP(k, k)A^T + Q \tag{19}$$

where Q is the state covariance matrix or process uncertainty. After the prediction step of the Kalman algorithm, a unit delay has to be applied to $\hat{x}(k+1, k)$ and $P(k+1, k)$ to obtain $\hat{x}(k, k-1)$ and $P(k, k-1)$ and evaluate the update equations, (15) to (17), in a new iteration cycle.

The Kalman filter is initialized with $\hat{x}(0) = 0$, since deviation variables are considered, and the estimation covariance matrix $P(0)$ with all its elements equal to 1000. The parameters Q and R are constant in this work. These parameters can be arbitrarily predefined: the process noise matrix Q with all its elements equal to 1000 and $R =$ 1000 (mg/dL)^2. However, these values can be tuned. The measurement noise R can be made equal to the square of the exactitude of the glucose sensor: for an exactitude of $\pm 15\%$ of reading [20], and a reference reading value of 119.5 mg/dL (steady initial BG value of the virtual patient), $R = 321$ (mg/dL)^2. The matrix parameter Q can be calculated using the state model evaluation with a PRBS for the EI rate input and the corresponding sample time. After collecting the EI input and state data, in deviation values, the process noise matrix Q can be adjusted as follows:

$$Q = \begin{bmatrix} \sigma_{x_1x_1} & \sigma_{x_1x_2} & \cdots & \sigma_{x_1x_m} \\ \sigma_{x_2x_1} & \sigma_{x_2x_2} & \cdots & \sigma_{x_2x_m} \\ \vdots & \vdots & \ddots & \vdots \\ \sigma_{x_mx_1} & \sigma_{x_mx_2} & \cdots & \sigma_{x_mx_m} \end{bmatrix} \tag{20}$$

where the elements $Q(i, j)$ correspond to variance $(i = j)$ or covariance $(i \neq j)$ values:

$$\sigma_{x_ix_j} = \frac{1}{N_{data} - 1} \sum_{l=1}^{N_{data}} \left(x_i(l) - \mu_{x_i}\right)\left(x_j(l) - \mu_{x_j}\right) \tag{21}$$

where μ_{x_i} is the mean value of state x_i, and N_{data} is the number of data.

2.5 *Performance Tests*

In order to evaluate the performance of the Kalman filter, the following tests are proposed: a glucose intake from a meal and the supply of an insulin dose.

Nutritional recommendations indicate that 53% of the total caloric content of a daily diet should come from carbohydrates, which are converted into glucose by the digestive process. The energy per gram of carbohydrate is 4 kcal/g [21]. For a 1500 kcal diet, a meal of 500 kcal should include 263 kcal from 66 g of carbohydrates. Therefore the glucose intake disturbance input to the physiological model is defined as a pulse with an amplitude of 2640 mg/min and a duration of 25 min. It is important to notice that the glucose input is not considered in the discrete linear model. An insulin dose of 225 mU is supplied with a low infusion rate of 15 mU/min for 15 min. This test is applied separately from the glucose intake simulation.

Additionally, the gain and time constant parameters of the BG sensor are proposed to change in ±60% of their nominal value, during the meal and EI tests, for further evaluation of the state estimation system. The loss of sampled data during 15, 30 and 60 min is also considered. The performance index is defined as the integral of the square BG estimation error (ISE) during the total simulation time (1800 min).

3 Results

Discrete Linear Model. The identification was performed using sample times of 1 and 15 min. The structure of the ARX model was changed to detect the minimum complexity required to approach the data from the virtual patient. Tables 1 and 2 list

Table 1 Mean square error (*mse*) for ARX(*na*, *nb*, *d*) models with $T = 1$ min

na, nb	$mse, d = 0$	$mse, d = 1$	$mse, d = 2$	$mse, d = 3$	$mse, d = 4$	$mse, d = 5$
2, 2	1.20E−3	3.83E−4	8.56E−4	1.40E−3	1.80E−3	1.8E−3
2, 3	3.10E−4	1.58E−4	8.37E−4	1.40E−3	1.80E−3	1.70E−3
3, 3	1.50E−5	1.10E−5	5.40E−5	5.29E−4	5.55E−4	5.74E−4
3, 4	6.22E−6	1.22E−5	5.61E−4	5.38E−4	5.27E−4	5.81E−4
4, 4	4.44E−6	1.31E−5	2.61E−4	6.49E−4	5.04E−4	4.88E−4
4, 5	4.64E−6	1.52E−5	2.55E−4	4.31E−4	4.90E−4	4.90E−4

Table 2 Mean square error (*mse*) for ARX(*na*, *nb*, *d*) models with $T = 15$ min

na, nb	$mse, d = 0$	$mse, d = 1$	$mse, d = 2$	$mse, d = 3$	$mse, d = 4$	$mse, d = 5$
2, 2	0.6045	4.1239	4.0116	5.1684	5.4488	5.5713
2, 3	0.2849	4.1833	3.2298	5.2423	5.5723	3.8697
3, 3	0.1686	4.1024	3.0763	4.0741	5.5347	3.8473
3, 4	0.1433	3.0733	3.0508	3.8090	3.8217	3.1701
4, 4	0.1664	2.8680	2.7385	3.2489	2.5583	2.9243
4, 5	0.0932	2.8259	2.4289	2.5481	2.4822	2.6789

the mse obtained for different structures. Increasing the delay parameter d has more effect on reducing the mse for low-order models. The number of coefficients na and nb can clearly reduce the mse more effectively than parameter d. However, a higher complexity of the structure is not justified by a negligible improvement of the mse. The chosen structures for both sample times were ARX(3, 3, 0):

$$\theta = [2.5915, -2.2174, 0.6253, -0.0002, -0.0017, -0.0009]', \quad \text{for T} = 1 \text{ min} \tag{22}$$

$$\theta = [1.3417, -0.5661, 0.1137, -0.4507, -0.2571, 0.2510]', \quad \text{for T} = 15 \text{ min} \tag{23}$$

Kalman Filter Performance. The Kalman filter was used with the predefined parameters for the process and measurement uncertainties of $Q = 1000J_3$ and $R = 1000$, respectively, as well as with the calculated parameters, as explained in Sect. 2. The process noise characterization resulted in

$$Q = 10^7 \begin{bmatrix} 1.1866 & 1.1757 & 1.1639 \\ 1.1757 & 1.1652 & 1.1537 \\ 1.1639 & 1.1537 & 1.1427 \end{bmatrix}, \quad \text{for T} = 1 \text{ min} \tag{24}$$

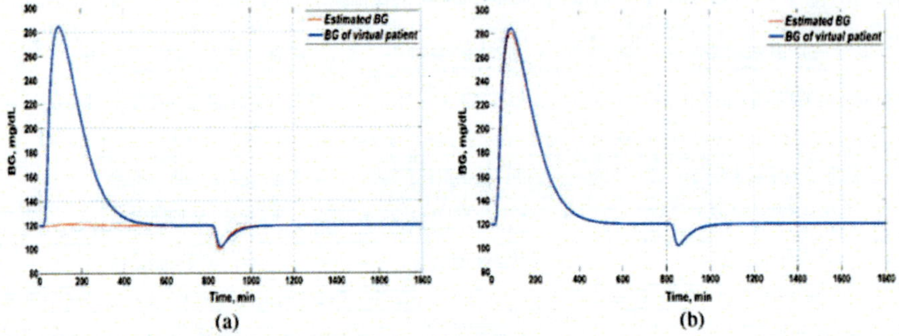

Fig. 3 Kalman filter with T = 1 min: **a** use of predefined parameters, ISE = 3,226,639.54; **b** use of calculated parameters ISE = 4607.36

$$Q = \begin{bmatrix} 935.0103 & 916.9251 & 878.5736 \\ 916.9251 & 934.9365 & 916.8578 \\ 878.5736 & 916.8578 & 934.8759 \end{bmatrix}, \quad \text{for T} = 15 \text{ min} \tag{25}$$

The estimated BG is compared against the virtual patient's BG in Fig. 3 for T = 1 min, and in Fig. 4 for T = 15 min. The first part of the graphs (from time 0 to 800 min) shows the response to the meal test, and the second part (from time 800 to 1800 min) corresponds to the BG reaction to the EI test. The ISE indicator is evaluated over the entire and fixed simulation time.

The tuning of parameters Q and R results in a noticeable reduction of the estimation error. The ISE decreases in 99% for T = 1 min and 60% for T = 15 min (Figs. 3 and 4).

The Kalman filters with adjusted uncertainty parameters were further tested considering random variations of the sensor dynamics during the meal and EI supply. Figure 5 shows the response with different sample times, when the gain and time constant of the sensor system change in ±60%.

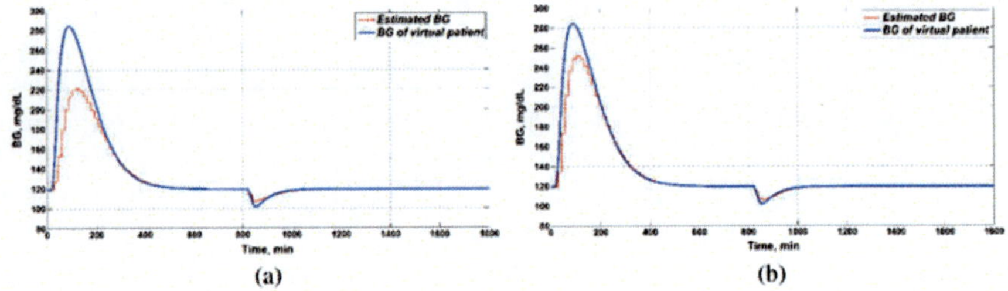

Fig. 4 Kalman filter with T = 15 min: **a** use of predefined parameters with ISE = 598,007.71; **b** use of calculated parameters; ISE = 235,520.41

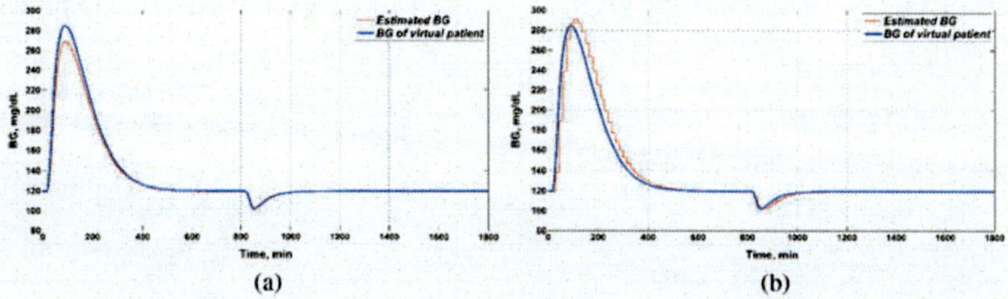

Fig. 5 Kalman filter estimation with ±60% random variation of gain and time constant of the BG sensor: **a** T = 1 min, ISE = 88,465.83; **b** T = 15 min, ISE = 362,816.90

The performance observed in Fig. 5 is deteriorated as shown by the ISE increment of 1820% for T = 1 min, and 54% for T = 15 min, with respect to the scenario with constant sensor dynamics (Figs. 3b and 4b).

Table 3 summarizes the effect of using calculated process and measurement noise parameters in the Kalman filter. The estimation error with adjusted Q and R is less than the obtained error with predefined values even with the changing dynamics of the BG meter.

Another aspect to consider is the loss of information from the sensor, due to malfunction of the device or to the voluntary interruption of BG measurement. When a BG measurement is missing, no residue or innovation can be quantified and the second term of the update Eq. (16) is suppressed. BG sample loss for 15, 30 and 60 min was simulated at the same time that random variations of the sensor dynamics were applied. Table 4 shows ISE values obtained with the tuned Kalman filter. Figure 6 shows that the Kalman filter tracks BG tendency although with low exactitude.

Table 3 Comparison of ISE indicators when Q and R are calculated and measurement dynamics changes (variations of ±60% of nominal gain and time constant)

Sample time, T (min)	ISE Predefined Q and R; constant meter dynamics	ISE Calculated Q and R; constant meter dynamics	ISE Calculated Q and R; variable meter dynamics
1	3,226,639.54	4607.36	88,465.83
15	598,007.71	235,520.41	362,816.90

Table 4 Comparison of ISE indicators when BG measurement signal is lost for a specific period

Sample time, T (min)	ISE No signal loss	ISE 15 min loss	ISE 30 min loss	ISE 60 min loss
1	88,465.83	174,195.69	632,746.24	834,063.64
15	362,816.90	417,148.40	512,648.10	1,305,177.46

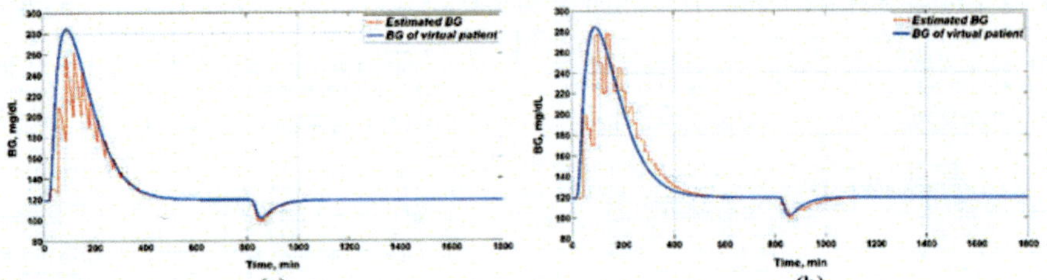

(a) (b)

Fig. 6 Kalman filter estimation with missing sampled BG data during 30 min: **a** T = 1 min, ISE = 632,746.24; **b** T = 15 min, ISE = 512,648.10

4 Conclusions

The contribution of this work is the stochastic identification of the glucose-insulin process and the use of the identified model to implement the Kalman filter. The stochastic identification allowed the detection of third-order linear models as adequate for representing the process behavior, and implementing support discrete systems. The Kalman filter implementation used the resulting models as the basis for the prediction phase, and also as a resource to tune parameters of the update phase in order to achieve better BG estimation.

The calculation of the uncertainty parameters may sometimes be neglected when using the Kalman filter, opting to use reasonable default values. However, this study shows that proper tuning of such parameters is important for critical applications such as the monitoring of the condition of a diabetic patient. In the presented simulations, the estimation error with adjusted Q and R was clearly lower than the error obtained with predefined values, even with the changing dynamics of the BG meter.

The estimation with the Kalman filter was tested when BG measurements were lost for a specific period. This extreme situation for the evaluation of the filter, may not be unusual for a patient under continuous monitoring when it is necessary to remove the sensor temporarily. Although an increased error is obtained, the Kalman filter keeps estimating BG level, which would allow continuity in the treatment or alerting on the trend and direction of change of BG.

The single input of the model for EI is another restriction of the model, besides the reduced third order and the linearity, that explains the lower accuracy in the estimation during the glucose intake test. However, the adjustment of the filter parameters was useful for handling unconsidered inputs. This observation indicates that the Kalman filter based on a simplified model is useful even for unmodeled disturbances.

The Kalman filter also generates the evaluation of the state variables. The exactitude of the BG output estimation reflects the adequacy of the state estimation. In the case of the linear reduced order model approximated to the glucose-insulin dynamics, the states may not represent specific physiological variables, but an abstract combination of them. However, the few states that define the discrete model provide a good basis for a control strategy. Future work is directed to taking full advantage of the discrete Kalman filter by using the state variables in the design of a control law to reduce BG deviations and stabilization times before different disturbances.

References

1. Asif M (2014) The prevention and control the type-2 diabetes by changing lifestyle and dietary pattern. J Educ Health Promot 3:1. https://doi.org/10.4103/2277-9531.127541
2. Islam MS (2021) Diabetes: from research to clinical practice. Adv Exp Med Biol 1307:1–5. https://doi.org/10.1007/5584_2020_553
3. International Diabetes Federation (2019) IDF diabetes atlas, 9th edn. Brussels. Available at: https://www.diabetesatlas.org
4. Sorensen JT (1985) A physiologic model of glucose metabolism in man and its use to design and assess improved insulin therapies for diabetes. Thesis, Massachusetts Institute of Technology
5. Lema-Perez L, Aguirre-Zapata E, Garcia-Tirado J (2015) Recent advances in mathematical models for the understanding and treatment of type 1 diabetes mellitus. In: 2015 IEEE 2nd Colombian conference on automatic control (CCAC), Manizales, pp 1–6. https://doi.org/10.1109/CCAC.2015.7345212
6. Bondia J, Romero-Vivo S, Ricarte B, Diez JL (2018) Insulin estimation and prediction: a review of the estimation and prediction of subcutaneous insulin pharmacokinetics in closed-loop glucose control. IEEE Control Syst 38(1):47–66. https://doi.org/10.1109/MCS.2017.2766312
7. Patra AK, Nanda A (2020) An automatic insulin infusion system based on Kalman filtering model predictive control technique. J Dyn Syst Meas Control 143(2):021004 (11 pages). https://doi.org/10.1115/1.4048370
8. Nanda A, Patra AK (2020) Kalman filtering linear quadratic regulator for artificial pancreas in type-I diabetes patient. Int J Model Identif Control 34(1):59–74. https://doi.org/10.1504/IJMIC.2020.108916
9. Batora V, Tarnik M, Murgas J, Schmidt S, Norgaard K, Poulsen NK, Madsen H, Jorgensen JB (2015) Bihormonal control of blood glucose in people with type 1 diabetes. In: 2015 European control conference (ECC), Linz, pp 25–30. https://doi.org/10.1109/ECC.2015.7330520
10. Facchinetti A, Sparacino G, Cobelli C (2010) Enhanced accuracy of continuous glucose monitoring by online extended Kalman filtering. Diabetes Technol Ther 12(5):353–363. https://doi.org/10.1089/dia.2009.0158
11. Sparacino G, Facchinetti A, Cobelli C (2010) 'Smart' continuous glucose monitoring sensors: on-line signal processing issues. Sensors 10(7):6751–6772. https://doi.org/10.3390/s100706751
12. Hou L, Zhang H, Wang J, Shi D (2019) Optimal blood glucose prediction based on intermittent data from wearable glucose monitoring sensors. In: 2019 Chinese control conference (CCC), Guangzhou, pp 5463–5467. https://doi.org/10.23919/ChiCC.2019.8866572
13. Eren-Oruklu M, Cinar A, Rollins DK, Quinn L (2012) Adaptive system identification for estimating future glucose concentrations and hypoglycemia alarms. Automatica 48(8):1892–1897. https://doi.org/10.1016/j.automatica.2012.05.076
14. Yang J, Li L, Shi Y, Xie X (2019) An ARIMA model with adaptive orders for predicting blood glucose concentrations and hypoglycemia. IEEE J Biomed Health Inform 23(3):1251–1260. https://doi.org/10.1109/JBHI.2018.2840690
15. Bououden S (2016) High gain observer design for blood glucose in type I diabetics. In: 2016 8th international conference on modelling, identification and control (ICMIC), Algiers, pp 643–647. https://doi.org/10.1109/ICMIC.2016.7804190
16. American Diabetes Association (2020) 2. Classification and diagnosis of diabetes: standards of medical care in diabetes—2020. Dia Care 43(Supplement 1):S14–S31. https://doi.org/10.2337/dc20-S002

17. Profil (2015) Hyperinsulinemic euglycemic clamp at Profil. Available at: https://www.profil.com/services/hyperinsulinemic-euglycemic-clamp. Accessed 24 Mar 2021
18. Okawara H (2013) DSP-based testing—fundamentals 50 PRBS (pseudo random binary sequence). Advantest. Available at: https://www3.advantest.com/documents/11348/3e95df23-22f5-441e-8598-f1d99c2382cb. Accessed 24 Mar 2021
19. Haykin S (2014) Adaptive filter theory, 5th edn. Prentice Hall, New Jersey
20. Freckmann G, Link M, Schmid C, Pleus S, Baumstark A, Haug C (2015) System accuracy evaluation of different blood glucose monitoring systems following ISO 15197:2013 by using two different comparison methods. Diabetes Technol Ther 17(9):635–648. https://doi.org/10.1089/dia.2015.0085
21. Whitney EN, Rolfes SR (2016) Understanding nutrition. Cengage Learning, Stamford, CT

Simultaneous Optimization in a Mill for Juice Extraction of Sugar Cane and Agave

Armando Mares Castro

Abstract This article presents a case of simultaneous optimization of two variables of interest in a mill prototype for juice extraction process in sugar cane and agave tequilana weber. Response surface models and optimization methods by desirability function are applied to obtain the best combination of operating parameters in a mill prototype to maximize extraction efficiency and reduce amperage consumption. The experimentation was carried out in a prototype mill, using sugar cane in the experimental tests for the extraction of the juice. The results obtained were satisfactory, improving extraction efficiency by 18% and reducing consumption by 5%.

Keywords Desirability function · Response surface methodology · Experimentation · Juice extraction

1 Introduction

Continuous improvement involves the analysis of the process conditions that allow the best operating conditions in the products and processes. The desired quality characteristics, also called responses, are subject to measurement, based on this information, better operating conditions will be sought to optimize the process. The optimization of the process implies obtaining the highest levels of the variables (characteristics class *bigger is the best*) the lowest levels (characteristics class *smaller is better*) or subject to a nominal value (characteristics class *Nominal is the best*). There are situations in which it is sought to optimize several quality characteristics that can seek different objectives, the solution found must satisfy in the best way all the objectives at the same time. In multi-objective optimization, a solution that satisfies multiple objectives will generally represent the determination of a trade-off

A. Mares Castro (✉)
Tecnológico Nacional de México/ITS de Purísima del Rincón, Blvd. Del Valle # 2301, Guardarrayas, Purísima del Rincón, Guanajuato C.P. 36413, México
e-mail: armando.mc@purisima.tecnm.mx

K. L. Flores Rodríguez et al. (eds.), *Recent Trends in Sustainable Engineering*, Lecture Notes in Networks and Systems 297,
https://doi.org/10.1007/978-3-030-82064-0_11

that can work to the advantage of some variables and reduce it in others. Based on this approach, it is possible to determine the best solution based on predefined criteria.

This article presents a proposal for selecting the best solution for a multi-response optimization problem. The modeling of the variables is carried out with the response surface methodology (RSM) in order to obtain the best combination of parameters that maximize the juice extraction efficiency in a mill and in to reduce the amperage consumption simultaneously. The efficiency of the mill is evaluated based on the maximization of the extraction of juice, which is made by grinding process for sugar and/or tequila production process with agave tequilana weber. Experimental tests were made in a scale prototype mill and sugar cane was used in the experimental runs.

Once the optimal parameters have been obtained in the prototype, the optimal parameters can be scaled to a bigger machine currently operating in the tequila factory. In the literature there are some cases of optimization in the agave juice extraction process, but the case presented is original, since the experimental parameters are related to the design and elements of the machine. Another contribution is the comparison of the efficiency of multi-response optimization techniques in a beverage manufacturing process.

2 Multi-objective Optimization and RSM

Through the use of an experimental scheme it is possible to model the responses of interest through the use of regression models. In the RSM, first and second order models can be used, for the second case, the most used designs with the central compound design and the Box-Behnken design, their properties have been published in specialized books on the method [1–3]. In the RSM for r answers there are r models, the jth model for the answer Y_i are written as in Eq. 1.

$$Y_i = \beta_{i0} + \mathbf{X}'\beta_i + \mathbf{X}'\mathbf{B}_i\mathbf{X} + \varepsilon_i \tag{1}$$

where $\mathbf{X}' = (x_1, \ldots, x_k)$ k factors, β_{i0} the constant, $\beta_i = (\beta_{i1}, \ldots, \beta_{ik})$ simmetric matrix of second order parameters, and $\varepsilon_i \sim N\left(0, \sigma_i^2\right),\ j = 1, \ldots, r$. The estimation of the parameters that characterize each of the models for the responses are obtained through statistical regression techniques. The classical assumptions must be fulfilled regarding the structure of the experimental data, such as normality, homoscedasticity and independence between the observations of the residuals.

2.1 The Desirability Function

This multi-response optimization technique was developed by Harrington [4] and later modified by Derringer and Suich [5]. In this method, an individual desirability function is defined for each of the response variables and, based on them, a global desirability is defined. The desirability function has been widely used in industrial and manufacturing processes, such as chemical processes [6], materials [7–10], machining [11, 12], food [13], among other áreas of research. Desirability $d_j (j = 1, \ldots, r)$ is defined in Eq. 2 for a problem class (a) *Nominal is the best*, (b) *Bigger is the best* and (c) *Smaller is the best*.

$$
\begin{aligned}
&(a)\quad d_j\left(\widehat{Y}_j(x)\right) = \begin{cases} 0 & \textit{if } \widehat{Y}_j(x_i) \le \widehat{Y}_j^- \textit{ or } \widehat{Y}_j(x_i) \ge \widehat{Y}_j^+ \\ 1 - \frac{M_j - \widehat{Y}_j(x_i)}{M_j - \widehat{Y}_j^-} & \textit{if } Y_j^- < \widehat{Y}_j(x) \le M_j \\ 1 - \frac{\widehat{Y}_j(x_i) - M_j}{\widehat{Y}_j^+ - M_j} & \textit{if } M_j \le \widehat{Y}_j(x) < \widehat{Y}_j^+ \end{cases} \\
&(b)\quad d_j\left(\widehat{Y}_j(x)\right) = \begin{cases} 0 & \textit{if } \widehat{Y}_j(x_i) \le \widehat{Y}_j^- \\ 1 - \frac{M_j - \widehat{Y}_j(x_i)}{M_j - \widehat{Y}_j^-} & \textit{if } Y_j^- < \widehat{Y}_j(x) < M_j \\ 1 & \textit{if } \widehat{Y}_j(x_i) \ge M_j \end{cases} \\
&(c)\quad d_j\left(\widehat{Y}_j(x)\right) = \begin{cases} 0 & \textit{if } \widehat{Y}_j(x_i) \ge \widehat{Y}_j^+ \\ 1 - \frac{M_j - \widehat{Y}_j(x_i)}{M_j - \widehat{Y}_j^+} & \textit{if } M_j < \widehat{Y}_j(x) < \widehat{Y}_j^+ \\ 1 & \textit{if } \widehat{Y}_j(x_i) \le M_j \end{cases}
\end{aligned}
\tag{2}
$$

For the *Nominal is the best* case, when value of $\widehat{Y}_j(x_i) \le \widehat{Y}_j^-$ *or* $\widehat{Y}_j(x_i) \ge \widehat{Y}_j^+$, the desirability function has a value of 0 and indicates that there is no conformity with the desired goal, and when $\widehat{Y}_j(x) = M_j$ the best possible value is reached and has total conformity, that is: $d_j(M_j) = 1$. The desirability function depends on the conditions of the process and therefore it may also be desired to minimize or maximize the response, in the first case $M_j = Y_j^-$ is substituted in the third function of expression 3 and in the other case $M_j = Y_j^+$ goes in the second function of Eq. 2. Dimensions Y_j^- and Y_j^+ in each of the responses must be specified to define the desirability function. These can be based on the specification limits of the products or processes, on company standards or simply on the experimenter's judgment. They can also be determined in terms of the physical ranges of the response and in this case they are presented as the extreme values of each response: $Y_j^- = \min\left(\widehat{Y}_j(x)\right)$ and $Y_j^+ = \max\left(\widehat{Y}_j(x)\right)$. The individual desirabilities are combined using the weighted geometric mean to obtain an overall desirability in Eq. 3

$$
D(x) = \left(d_1(x)^{w_1} \times d_2(x)^{w_2} \times \ldots \times d_j(x)^{w_j}\right)^{1/\sum w_j} \tag{3}
$$

where the weights w_j are constants that balance the relative importance of each variable.

3 Materials and Methods

The steps for the proposed methodology are illustrated below:

Step 1: Analysis of the process and determination of the experimental variables is performed.
Step 2: Selection of the experimental matrix and determination of the levels for the control factors.
Step 3: Conducting experimental runs.
Step 4: Analysis of variance, determination of regression models and verifying the assumptions for the residuals.
Step 5: Determination of the optimization scheme.
Step 6: Multi-objective optimization by desirability function.
Step 7: Selection of the best parameters for the process.

Analysis of the process The grinding process has historically been carried out in different ways, from the use of wood, steel mallets, circular stones of more than one meter in diameter and 50 cm thick around a circular pool. The resulting juice was collected by hand with the use of containers. After the 1950s, mills began to be used in which the agave passed through a cutter, currently, highly automated systems are designed from the agave harvest [14]. The prototype used for the experimentation is shown in Fig. 1.

The grinding process has a leftover bagasse that can be used in the textile sector [15, 16]. There is research available on optimization of parameters in the tequila

Fig. 1 Mill prototype. *Source* Authors

Fig. 2 Amperage indicators. *Source* Authors

production process [17], as well as in the extraction process [18]. In these investigations, an approach is used to manipulate variables present in the process or chemical elements. In this research, mill variables are used in order to improve grinding efficiency (improve juice extraction) and reduce amperage consumption in roller motors when grinding. The machine has three amperage indicators (Fig. 2), one for each motor, which transmit the turning movement to the rollers that make up the grinding system. For the measured amperage response, the average amperage consumption at the time of the grinding process will be handled with a simple average for each experimental run as indicated in Eq. 4.

$$\overline{x}_A = \frac{\sum_{i=1}^{3} A_i}{n} = \frac{A_1 + A_2 + A_3}{3} \tag{4}$$

The control factors selected for the experiment are: hydraulic pressure x_1, speed of the rollers x_2 and opening of the rollers x_3. The rollers and its opening are shown in Fig. 3.

The other response variable of the experiment is related to the efficiency of juice extraction. For the experimental tests, sugar canes were used, which have a behavior similar to agave when grinding. To evaluate this variable, canes of the same initial

Fig. 3 Roller opening. *Source* Authors

Fig. 4 Initial weight of the sugar cane. *Source* Authors

weight (1475 kg) were cut (Fig. 4), then ground and subsequently the weight was measured again as shown in Eq. 6. The subtraction is the weight of juice obtained by grinding, with which has a variable of the *bigger is better* type.

$$w_{juice} = initial\_w_{cane} - final\_w_{cane} \tag{5}$$

3.1 Levels and Experimental Matrix

The experimental levels for the control factors were determined based on the capabilities of the prototype. 3 levels were used for each of the control variables, low level, medium level and high level, which are shown in Table 1.

Experimental matrix and responses obtained are shown in Table 2. A Box-Behnken experimental design with 15 experimental runs was used. The response Efficiency in the extraction of juice Y_1 was determined according to Eq. 6 and the response average amperage consumption Y_2 was determined by taking the measurements of the 3 indicators at the time of grinding in the experimental run, the data obtained is the average consumption indicated in Eq. 4.

Once the experimental data have been obtained, the corresponding statistical studies are carried out.

Table 1 Levels for the experimental factors

Level	Coded	x_1 (MPa)	x_2 (rpm)	x_3 (mm)
Low	−1	4.14	3.50	55.0
Mid	0	6.89	4.75	59.5
High	1	9.65	6.00	64.0

Source Authors

Table 2 Experimental matrix and responses obtained

Run	x_1	x_2	x_3	Y_1	Y_2
1	−1	−1	0	0.620	2.900
2	1	−1	0	0.670	3.033
3	−1	1	0	0.575	2.500
4	1	1	0	0.570	2.833
5	−1	0	−1	0.550	2.967
6	−1	0	1	0.880	3.167
7	1	0	−1	0.745	3.133
8	1	0	1	0.895	3.167
9	0	−1	−1	0.400	2.867
10	0	−1	1	0.840	2.900
11	0	1	−1	0.545	2.833
12	0	1	1	0.905	2.933
13	0	0	0	0.665	2.967
14	0	0	0	0.705	2.933
15	0	0	0	0.645	2.933

Source Authors

4 Results and Discussion

The first step in a multi-response optimization study is to analyze whether there is a correlation between the responses, since if there is, there may be difficulties in determining an efficient trade-off solution. Figure 5 shows the correlation matrix between variables Y_1 and Y_2, the value of the pair correlation for the variables is 0.556 and a p-value of 0.031, indicating a low correlation between the variables.

$$\begin{aligned}\widehat{Y}_1 = 0.6717 + 0.0319x_1 + 0.0081x_2 + 0.1600x_3 - 0.0137x_1x_2 \\ -0.0450x_1x_3 - 0.0200x_2x_3 + 0.0160x_1^2 - 0.0790x_2^2 + 0.0798x_3^2\end{aligned} \quad (6)$$

$$\begin{aligned}\widehat{Y}_2 = 2.9400 + 0.0790x_1 - 0.0751x_2 + 0.0459x_3 + 0.0500x_1x_2 \\ -0.0415x_1x_3 + 0.0167x_2x_3 + 0.0487x_1^2 - 0.1765x_2^2 + 0.1155x_3^2\end{aligned} \quad (7)$$

Table 3 shows the estimated coefficients and the adjustment statistics, the complete models are shown in Eqs. 6 and 7. Full models are used to improve the efficiency of the optimization method. Figures 6 and 7 show the graphs for the residuals in the models, a good compliance is observed for the assumptions of independence, normality and homoscedasticity in the residuals for the two variables.

Since one of the variables is *bigger is better* type and the other is *smaller is better* type, there are no references on a lower or upper limit for the process. Target values are defined at ±3 standard deviations from the mean and the mean is used as the lower limit for the case of the variable Y1 and as the upper limit for the variable Y_2

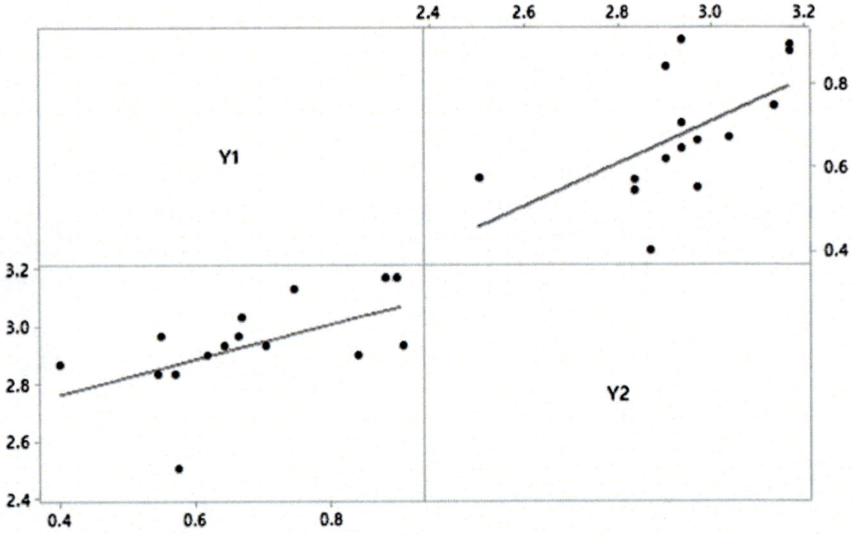

Fig. 5 Corelation matrix for Y_1 and Y_2. *Source* Authors

Table 3 Beta coefficients and fit statistics

Coef	Y_1	Y_2
$\widehat{\beta_0}$	0.6717	2.9400
$\widehat{\beta_1}$	0.0319	0.0790
$\widehat{\beta_2}$	0.0081	−0.0751
$\widehat{\beta_3}$	0.1600	0.0459
$\widehat{\beta_1}1$	0.0160	0.0487
$\widehat{\beta_2}2$	−0.0790	−0.1765
$\widehat{\beta_3}3$	0.0798	0.1155
$\widehat{\beta_1}2$	−0.0137	0.0500
$\widehat{\beta_1}3$	−0.0450	−0.0415
$\widehat{\beta_2}3$	−0.0200	0.0167
R^2	0.8906	0.8452
R^2_{adj}	0.6936	0.5664
p-value LoF	0.0816	0.0198
MSE	0.0068	0.0116
Std. dev.	0.0822	0.1078

Source Authors

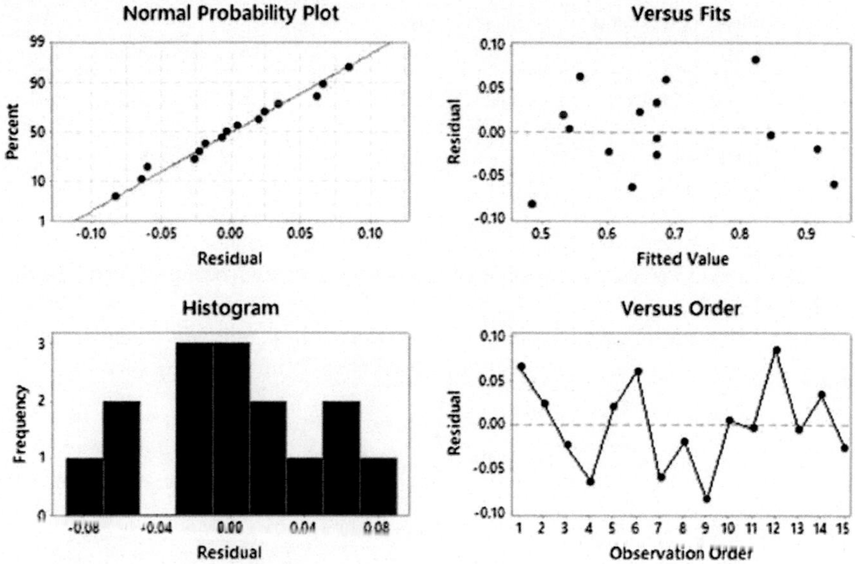

Fig. 6 Plots of residuals for $\widehat{Y}_1$. *Source* Authors

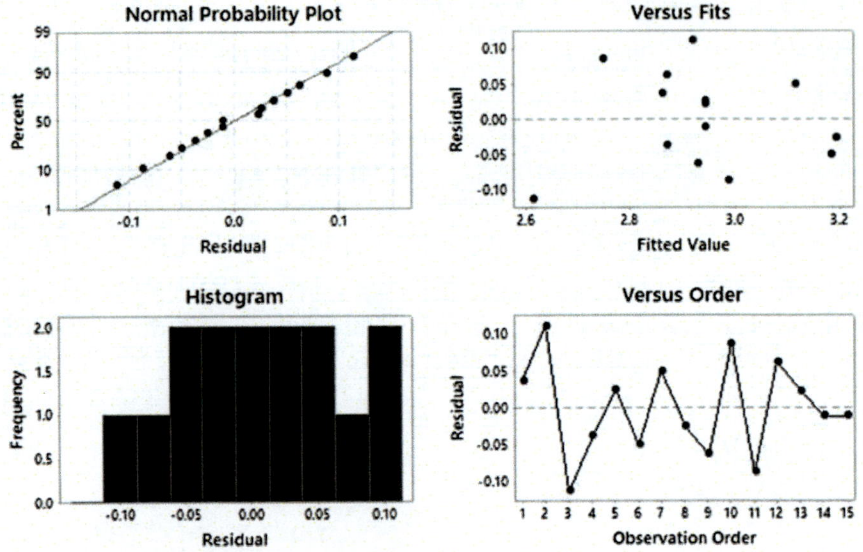

Fig. 7 Plots of residuals for $\widehat{Y}_2$. *Source* Authors

Table 4 Levels for the experimental factors

Spec	Y_1 maximize	Y_2 minimize
Lower limit	0.6807	
Target	0.9273	2.6166
Upper limit		2.9400

Source Authors

as shown in Table 4. These restrictions will be taken for resolution by the desirability function.

The optimization problem for the desirability function is shown in Eq. 8. The scheme is treated as a constrained nonlinear optimization problem.

$$
\begin{aligned}
&\text{Max}\left(\widehat{Y}_1(x_i)\right), \text{Min}\left(\widehat{Y}_2(x_i)\right)\\
&s.t.\\
&0.6807 \le \widehat{Y}_1(x_i) \le 0.9273\\
&2.6166 \le \widehat{Y}_2(x_i) \le 2.9400\\
&-1 \le x_i \le 1\\
&x_i \in R
\end{aligned} \tag{8}
$$

Figure 8 shows the overlay plots (graphical method), it is observed that a good solution is obtained by working x_1 at low level (-1, 4, 14 MPa of hydraulic pressure), x_2 at high level (1, 6 rpm of speed) and x_3 at high level (1.64 mm aperture).

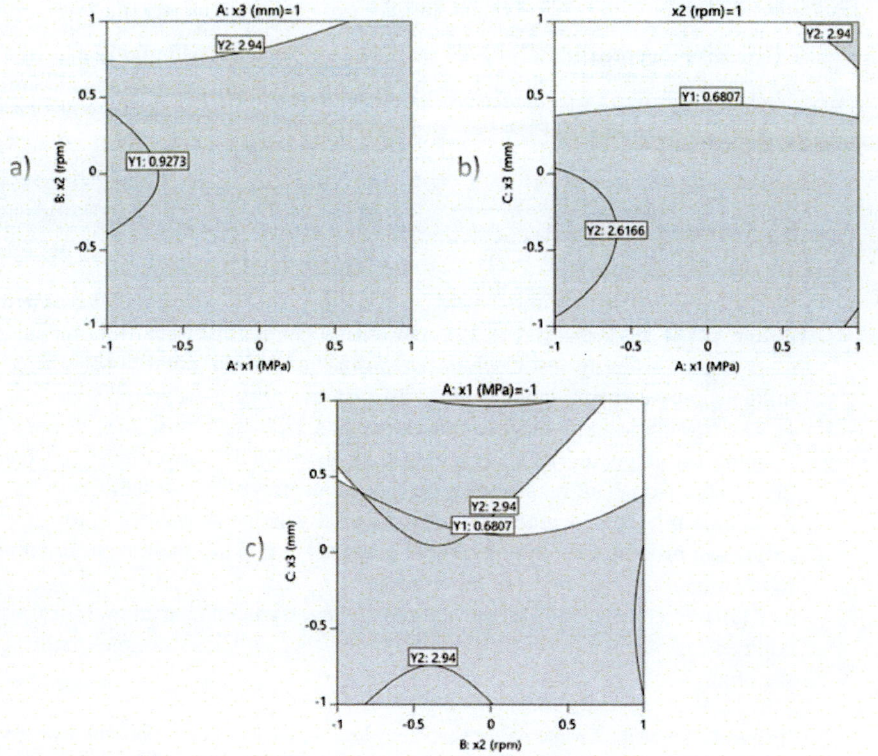

Fig. 8 Overlay plots for interactions **a** $x_1 - x_2$, **b** $x_1 - x_3$ and **c** $x_2 - x_3$. *Source* Authors

Table 5 Performance comparison original versus optimal levels

Factor	Pressure (MPa)	Roller speed (rpm)	Roller opening (mm)	Y_1	Y_2
Original	6.89	4.75	59.5	0.6717	2.94
Optimal	4.14	6	63.54	0.829	2.799
Difference				+18.97%	−4.79%

Source Authors

The results obtained by the desirability function method are $x_1 = -1, x_2 = 1$ and $x_3 = 0.898$ in levels coded with an overall desirability of 0.512. The corresponding actual values are pressure: 4.14 MPa, roller speed: 6 rpm, and roller opening: 63.54 mm. The resulting extraction efficiency is 0.829 kg and the average consumption is 2.799 A. Table 5 shows a comparison of the original levels compared to the optimal ones obtained.

It is observed that the optimal levels obtained can improve the juice extraction efficiency by approximately 18%, while the amperage consumption can be reduced by approximately 5% compared to the original levels for the control factors that had been considered by design.

5 Conclusions

Statistical experimentation should be considered as a fundamental element in the design phase of industrial equipment, since better results can be projected than mechanical design by itself when performing under real conditions, which will be influenced by control elements and noise.

The prototype used in this application is a 1:10 scale design of a machine that is being implemented in a very important tequila company in Mexico located in Tequila, Jalisco. The results obtained in the prototype are scalable and have good expectations of improvement in the productivity and performance of the agave juice extraction process, which is a critical process for making tequila and involves a lot of the manufacturing costs of the product.

Multi-objective optimization techniques represent a powerful tool for the process engineer in which a trade-off must be found that satisfies multiple quality goals in the process.

Acknowledgements Thanks for the support by *Instituto de Innovación, Ciencia y Emprendimiento para la Competitividad en Guanajuato "IDEA GTO"* obtained in the FINNOVATEG 2019 call for the development of this project.

References

1. Myers RH, Montgomery DC, Anderson-Cook CM (2016) Response surface methodology: process and product optimization using designed experiments. Wiley, Hoboken, NJ
2. Khuri AI, Cornell JA (1996) Response surfaces: designs and analyses. CRC Press, London
3. Box GEP, Draper NR (1987) Model building and response surface. Wiley, New York, NY
4. Harrington EC (1965) The desirability function. Ind Qual Control 21:494–498
5. Derringer G, Suich R (1980) Simultaneous optimization of several response variables. J Qual Technol 12(4):214–219. https://doi.org/10.1080/00224065.1980.11980968
6. Silva B, Gonzaga LV, Fett R, Oliveira AC (2019) Simplex-centroid design and Derringer's desirability function approach for simultaneous separation of phenolic compounds from Mimosa scabrella Bentham honeydew honeys by HPLC/DAD. J Chromatogr A 1585(25):182–191. https://doi.org/10.1016/j.chroma.2018.11.072
7. Kalantzi S, Kekos D, Mamma D (2019) Bioscouring of cotton fabrics by multienzyme combinations: application of Box-Behnken design and desirability function. Cellulose 26:2771–2790. https://doi.org/10.1007/s10570-019-02272-9
8. Simsek B, Uygunoglu T (2017) A full factorial-based desirability function approach to investigate optimal mixture ratio of polymer concrete. Polym Compos 39(9):3199–3211. https://doi.org/10.1002/pc.24330

9. Boyaci AI, Baynal K (2019) Optimisation of elastomeric bearings vulcanisation process using response surface methodology and desirability function approach. J Rubber Res 22(4):187–193. https://doi.org/10.1007/s42464-019-00027-6
10. Singh RK, Ganwar S, Singh DK (2020) Exploration of desirability function analysis-based particle swarm optimization for magnetic abrasive finishing of mild steel. J Adv Manuf Syst 19(4):607–627. https://doi.org/10.1142/S0219686720500298
11. Saidi R, Fathallah BB, Mabrouki T, Belhadi S, Yallese MA (2019) Modeling and optimization of the turning parameters of cobalt alloy (Stellite 6) based on RSM and desirability function. Int J Adv Manuf Technol 100:2945–2968. https://doi.org/10.1007/s00170-018-2816-x
12. Hazir E, Erdinler ES, Koc KH (2018) Optimization of CNC cutting parameters using design of experiment (DOE) and desirability function. J For Res 29:1423–1434. https://doi.org/10.1007/s11676-017-0555-8
13. Melnik AF, Shumetov VG, Kondrashin BS, Mikhaylov MR (2020) Use of Harrington's desirability function in wheat grain quality assessment. IOP Conf Ser Earth Environ Sci 422:1–9. https://doi.org/10.1088/1755-1315/422/1/012008
14. Saldaña N, Vaca CG, Hlawinska RS, Saldaña A, Ortega AF, Quiroz C, Cabrera JM, Garcíreaz A (2020) Propuesta de una cosechadora-desmenuzadora de Agave tequiliana Weber para la producción de bioetanol y revisión de los procedimientos existentes. Acta Univ 22(2):9–18. https://doi.org/10.15174/au.2012.361
15. Palacios H, Hernández JA, Esquivel M, Toriz G, Rojas OJ, Sulbarán BC (2019) Isolation and characterization of nanofibrillar cellulose from Agave tequilana Weber bagasse. Adv Mater Sci Eng 2019:7 pages. https://doi.org/10.1155/2019/1342547
16. Kanimozhi M, Vasugi N (2012) Characterization of Agave vera-cruz Mill leaf fiber for textile applications—an exploratory investigation. J Nat Fibers 9(4):219–228. https://doi.org/10.1080/15440478.2012.733526
17. Chávez A, López IG, Gomez JF, Luna G, Ortíz RI, Andrade I (2014) Optimization of Agave tequilana Weber var. Azul juice spray drying process. J Chem 2014:10 pages. https://doi.org/10.1155/2014/915941
18. Flores E, Salazar JA, Ramos EG (2015) Application of a Box-Behnken design for optimizing the extraction process of agave fructans (Agave tequilana Weber var. Azul). J Sci Food Agric 96(11):3860–3866. https://doi.org/10.1002/jsfa.7582

Assessment of Cadmium Sulfide Nanoparticles Synthesis by Cadmium-Tolerant Fungi

José Daniel Aguilar Loa, Abril Castellanos-Angeles, Luis Ángel García-Tejeda, Andrea Margarita Rivas-Castillo, and Norma Gabriela Rojas-Avelizapa

Abstract Due to optical properties, cadmium sulfide nanoparticles (CdSNP) are of interest for applications in several fields. The biosynthesis of this type of nanomaterials using the fungal biomass is a promising alternative because of the extracellular production of nanoparticles; however, the presence of cadmium could represent a challenge for microorganisms due to its well-known toxicity. Then, the present study was isolate fungi from environmental samples. Samples were obtained from polluted sites in different locations in Hidalgo State, Mexico. Firstly, fungi were isolated in culture media added with different cadmium concentrations to afterwards evaluate their ability to produce CdSNP. The synthesis was done using cadmium nitrate and elemental sulphur as raw materials using the fungal biomass of each isolate. Dynamic light scattering jointly with fluorescence and UV-Vis analysis were used to evidence the synthesis of CdSNP, which allowed to demonstrate the presence of nanoparticles with sizes below 100 nm, as well as absorption and emission bands characteristic of CdSNP at 310 and 515 nm, respectively. Thus, this study reports the ability of four cadmium-tolerant fungi to produce CdSNP.

Keywords Semiconductor nanoparticles · Cadmium sulfide · Biosynthesis · Biomass

1 Introduction

Semiconductor nanoparticles (NPs) are materials composed by metallic and non-metallic elements with high electrical stability and magnetic, thermal, chemical and optical properties [1, 2]. This type of NPs presents a wide variety of size and structures, made by different elements of the periodic table including II–VI, III–V and

J. D. A. Loa · A. Castellanos-Angeles · N. G. Rojas-Avelizapa (✉)
Department of Environmental Biotechnology, Centro de Investigación en Ciencia Aplicada y Tecnología Avanzada del Instituto Politécnico Nacional, Querétaro, Qro., México
e-mail: nrojasa@ipn.mx

L. Á. García-Tejeda · A. M. Rivas-Castillo
Universidad Tecnológica de la Zona Metropolitana del Valle de México, Tizayuca, Hgo., México

K. L. Flores Rodríguez et al. (eds.), *Recent Trends in Sustainable Engineering*, Lecture Notes in Networks and Systems 297,
https://doi.org/10.1007/978-3-030-82064-0_12

IV–VI groups, as well as different metallic elements [3]. These groups of elements include Zn, Hg and Cd as cations in combination with S, O, Te, and Se anions to form nanomaterials such as ZnO, ZnS, CdTe, and CdS [1, 4]. Cadmium sulfide semiconductor nanoparticles have gained importance in several fields due to their optical properties, which can be used in the coating of electrodes, solar cells, infrared detectors, environmental and biological sensors, as well in the construction of potentiometric and microscopic sensors [5, 6].

Methods for CdSNP synthesis (physical, chemical and biological) have both advantages and disadvantages [7]. Chemical and physical methods usually generate nanoparticles of high purity, homogeneity, reproducibility, and the processes tend to be short time [8]. Nevertheless, several times, it is necessary to use a stabilizing agent besides of the nanoparticles capping, using either organic or inorganic molecules to improve stability and biocompatibility [4]. In addition, these syntheses could be expensive, and generate significant pollutant emissions, harmful wastes, besides the high energy consumption of the process [4, 8]. Otherwise, biological methods use living organisms such as plants, bacteria and fungi, as well as their organic material to synthetize and capping the nanoparticles [9] resulting nanomaterials with colloidal stability and biocompatibility. Furthermore, this kind of methods have the advantage of using non high purity materials, and to generate less environmental wastes, due the process is carried out by enzymatic reactions and chemical interactions between reagents and biomolecules from organisms [10–12].

In most cases, the purification of biogenic nanoparticles is often complicated, depending on the type of organism used for their synthesis. For bacteria and plants, it is necessary perform several steps to obtain and purify the NPs, such as the lysis of the cell wall or the extraction and purification of polar compounds, respectively [8]. On the other hand, nanoparticle syntheses by fungi are usually less complex due to the proteins, enzymes, compounds and molecules involved during the process can be found in the biomass or the cell free filtrate [10, 13, 14].

Fungi CdSNP biosynthesis has been investigated in intracellular and extracellular processes [15]. In the first case, *Schizosaccharomyces pombe* and *Candida glabrata* yeast-like are found [16, 17], while in the second case filamentous fungi as *Pleurotus ostreatus* [18], *Phanerochaete chrysosporium* [19], *Coriolus versicolor* [20], *Trichoderma harzianum* [21], *Aspergillus versicolor* and *Fusarium oxysporum* [7, 10] have been reported.

The mechanisms of CdSNP biosynthesis and capping process in fungi are not well understood. However, it is recognized that cadmium-tolerant fungi respond to cellular stress caused by the presence of cadmium ions with processes like absorption, secretion, precipitation and conjugation of metallic ions [9, 20]. These processes are responsible for the change in the metal solubility, besides, to reduce metallic ions to form the nanoparticles by nucleation. Active transporters, membrane proteins, reducing agents, metallothioneins, cysteine-riche molecules with sulfhydryl groups, and reductases enzymes are some cellular compounds responsible for those processes [20, 22]. For previous reasons, cadmium-tolerant fungi are promising organisms for the synthesis of CdS nanoparticles, since due to the adaptation to adverse conditions, they have developed the cellular processes that favor the expression of biomolecules involved in nanoparticles formation. Therefore, the goal of the present study was to obtain and evaluate the capacity of biogenic CdSNP production by the biomass of cadmium-tolerant fungi isolated from metal-polluted sites.

2 Materials and Methods

2.1 Isolation and Purification of Cadmium-Tolerant Fungi

Fungi were isolated from water, soil and sludge samples that came from different locations in the Hidalgo State, México. Isolation was performed on dextrose potato agar (PDA) media by plate spreading, using 10^{-3}, 10^{-5}, and 10^{-6} serial dilutions of samples with NaCl 0.85% (w/v), after 48 h of incubation at 28 °C.

Fungal colonies well separated with different morphology were evaluated for their tolerance to different cadmium salt ($CdSO_4*8H_2O$) concentrations (2.5, 10, 100, 500 and 1000 ppm) in PDA plates added with rose Bengal 0.01% (w/v) then incubated at 28 °C until clear growth [23]. Those fungal colonies able to grow at different concentrations of cadmium salt were purified through sequential passes to fresh media and then evaluated for their ability to produce CdSNP. For experimental, only those fungal isolates that growth at highest concentration were studied.

2.2 Morphological Identification of Cadmium-Tolerant Fungi

The fungi morphology was observed by bright field microscopy with the Axio Scope A1 microscope, Carl Zeiss at 40× with KOH 3% (w/v) staining. Once identified, only those distinct microorganisms and based on growth characteristics were selected and studied for their ability to produce nanoparticles.

2.3 Nanoparticles Synthesis

To obtain pre-inoculum, selected fungi were transferred and incubated on PDA in Petri dishes, at 28 °C for 10 days. Afterwards, 10 mm diameter mycelium disc per isolate was cut and used as inoculum for biomass production in 250 mL of MGYP media (g/L: 10, glucose; 6, malt extract; 5, casein peptone; 3, yeast extract) as follows [23, 24]; cultures were incubated at 28 °C, 150 rpm during 4 days. Then, the whole content of flask was vacuum filtered by sterile filter paper of fine-pore; and wet weight of the fungal biomass was determined after three washes with 50 mL of sterile distilled water [23]. Subsequently, 6.2 g of fungal biomass was deposited in flasks with S° 0.3 M and 62.5 mL of $Cd(NO_3)_2*4H_2O$ 1 mM to evaluate CdSNP production. Using 6.2 g of fungal biomass in 62.5 mL of sterile water, biological control was prepared, while inorganic control was obtained from S° 0.3 M, and 62.5 mL of $Cd(NO_3)_2*4H_2O$ 1 mM. Controls and CdSNP reaction were incubated at 30 °C and 150 rpm for 24 h. Finally, supernatant was recovered by vacuum filtration as described above, and passed through 0.20 μm sterile filters Sarstedt [24].

2.4 Nanoparticles Characterization

Characterization of CdSNP was performed by UV-Visible Genesys 10S spectrophotometer (Thermo Scientific) with an absorption wavelength scanning of 200–500 nm [13]. The fluorescence spectrophotometry analysis was by scanning within 400–650 nm with spectrum readings every 1 nm, photomultiplication (PMT) of 5 with an excitation wavelength of 365 nm, and 10.0 nm of emission slit (F96Pro fluorescence spectrophotometer, Luzeren). The hydrodynamic radius and polydispersity index of CdSNP samples were done using Dynamic Light Scatter (DLS) technique in the particle analyzer Litesizer 500 (Anton Paar) at 25 °C in water with a refractive index of 1.33, and viscosity of 0.89 mPa.s.

3 Results and Discussion

3.1 Fungal Isolation and Identification

Seventeen cadmium-tolerant microorganisms were obtained from environmental samples and isolated incubating them in presence of cadmium sulfate. As can be observed in Table 1; 64.7% of fungal isolates were obtained from water, 11.8% from soil and 23.5% from sludge samples. The isolates with the highest cadmium tolerance were CAS1 and ZML5 obtained from sludge and water samples, respectively, since they were able to grow up to 1000 and 500 ppm of $CdSO_4*8H_2O$, respectively. In addition, about 47% of fungal isolates were tolerant to 100 ppm of cadmium salt, of which, most of them came from water samples. Isolates coded as ZML4, MMS1, CAS1, and CAS2 were morphologically identified as yeast, showing tolerance to high cadmium concentrations (>100 ppm), possibly due to their ability to accumulate Cd ions intracellularly [25]; the remaining 13 isolates presented a morphology of filamentous fungi.

A preliminary selection of ten filamentous fungi was done based on their easy manipulation in liquid and solid culture, and their capability to produce large amounts of biomass, which is necessary for nanoparticles synthesis [26]. Isolates coded as MML1, MML2, and CAL1 were discarded since presented low biomass production; did not growth in MGYP and PDA medium, as well as presented high sensitivity to

Table 1 Fungal growth at different concentrations of cadmium sulfate

Isolates code	Sample	$CdSO_4*8H_2O$ (ppm)[a]	Isolates code	Sample	$CdSO_4*8H_2O$ (ppm)[a]
CAS1	Sludge	1000	CAL3	Water	100
ZML5	Water	500	CAS3	Sludge	10
CAS2	Sludge	100	CAS3.2	Sludge	10
ZML1	Water	100	MML1	Water	10
ZML2	Water	100	CAL1	Water	10
ZML4	Water	100	CAL2	Water	10
MML2	Water	100	MMS2	Soil	10
MML3	Water	100	ZML3	Water	2.5
MMS1	Soil	100			

[a]Maximum concentration of cadmium salt where isolate growth was observed

changes in temperature, thus the incubation time must be equal or lower than 4 days to speed up the process [23]. Also yeast were discarded due to in most cases, reports mention that the nanomaterials synthesis occurs intracellularly [27].

The wet biomass obtained for selected isolates was within 15 and 25 g per 250 mL. Isolates coded as CAS3.2, MML3 and ZML3 had the least variation in the total weight of biomass obtained (Fig. 1). Statistical data from Tukey's test indicated no significant difference between the average weight of wet biomass from isolates.

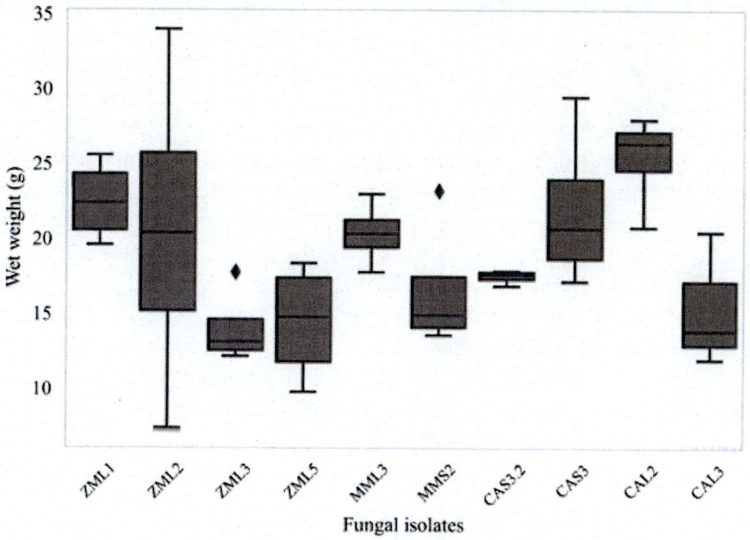

Fig. 1 Wet weight of fungal biomass per 250 mL of MGYP medium at 30 °C and 150 rpm for 24 h. Outliers over the boxes are indicated in rhombuses

3.2 UV-Vis Spectrophotometry

Data obtained from UV-Vis spectrophotometry showed that, negative inorganic control had a light absorption band at 225 nm, corresponding to cadmium nitrate, precursor of reaction [28]. Negative biological controls showed a band at 220 and 260 nm, possibly corresponding to biomolecules present in the medium, such as carbohydrates, polar compounds, amino acids and enzymatic cofactors as NADPH and NADH [9, 29, 30]. A peak at 330 nm for ZML1, MML3, MMS2, CAS3, CAS3.2, CAL2, and CAL3 samples was found, probably due to in response to stress, such as temperature, reactive oxygen species, light, and even by the sporulation process, secondary metabolites are released by fungi [31]. Last samples also showed absorption peaks at 230 and 260 nm. By the other hand, an absorption peak at 310 nm in CdSNP reactions synthesis of isolates ZML2, ZML3, and ZML5 was shown, which could be correspond to CdSNP production, as it was reported by authors [4, 16, 22, 24]. Fungal isolates coded as ZML1, MML3, CAS3, CAS3.2, CAL2, and CAL3 did not show any absorption band at 310 nm (Fig. 2).

3.3 Fluorescence Spectrophotometry

Fluorescence spectrophotometry for negative inorganic control showed a band at 415 nm from the Raman spectrum of water, which appears after 50 nm of the excitation wavelength [32]. Negative biological controls of fungal isolates showed fluorescence bands at 450 and 615 nm, corresponding probably to amino acids and proteins that can be excited from a range within 230 to 365 nm, and emit fluorescence bands ranging from 440 to 460 nm [33]. Target samples, where the reaction was carried out for CdSNP synthesis, showed a band approximately between 450 and 440 nm. The slight shift of these bands, regarding the negative biological control, could be due to the interaction of the cadmium and sulfur atoms as CdSNP with biomass [7, 24, 34]. CdSNP were evidenced with a fluorescence band around 515 nm (Fig. 3) as it has been reported in literature [20, 24, 35, 36]. The band at 515 nm for isolates ZML1, ZML5, MMS2 and CAS3.2 was greater (4 counts per second, cps), than for the rest of the fungal isolates. A band at 615 nm, in negative biological controls, also appeared for ZML5 and MMS2 isolates, which could be due to the presence of pigments and antioxidant molecules related to nucleophilic reactions used during the synthesis and capping of the CdSNP [37, 38].

3.4 Dynamic Light Scattering Analysis

Dynamic light scattering analysis for ZML1, ZML5, MMS2 and CAS3.2 isolates indicates that CdSNP in samples have an average hydrodynamic radius range from

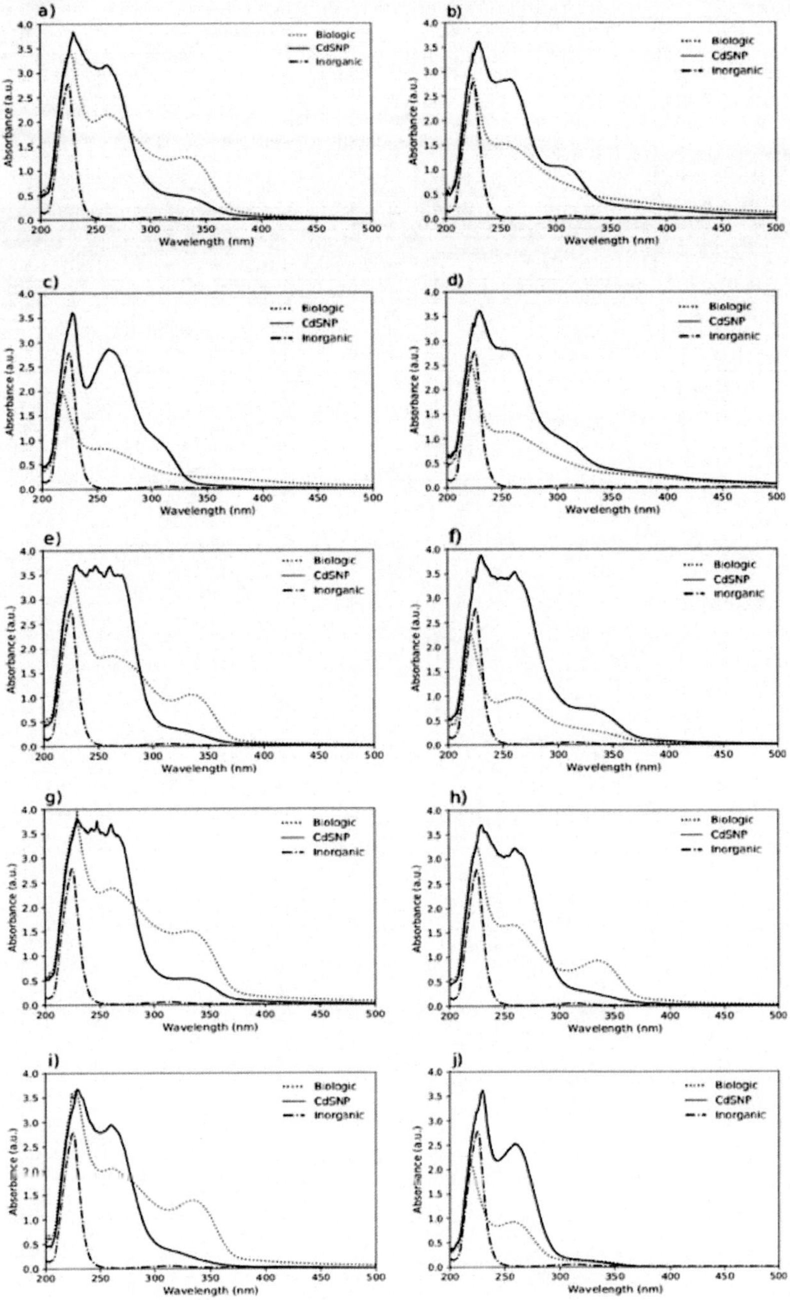

Fig. 2 UV-vis spectrophotometry analysis. Fungal isolates are indicated in different graphs: **a** ZML1; **b** ZML2; **c** ZML3; **d** ZML5; **e** MML3; **f** MMS2; **g** CAS3.2; **h** CAS3; **i** CAL2; and **j** CAL3. Negative biologic and inorganic controls are coded as "Biologic" and "Inorganic", whereas the sample where CdSNP synthesis reaction was done is indicated as "CdSNP"

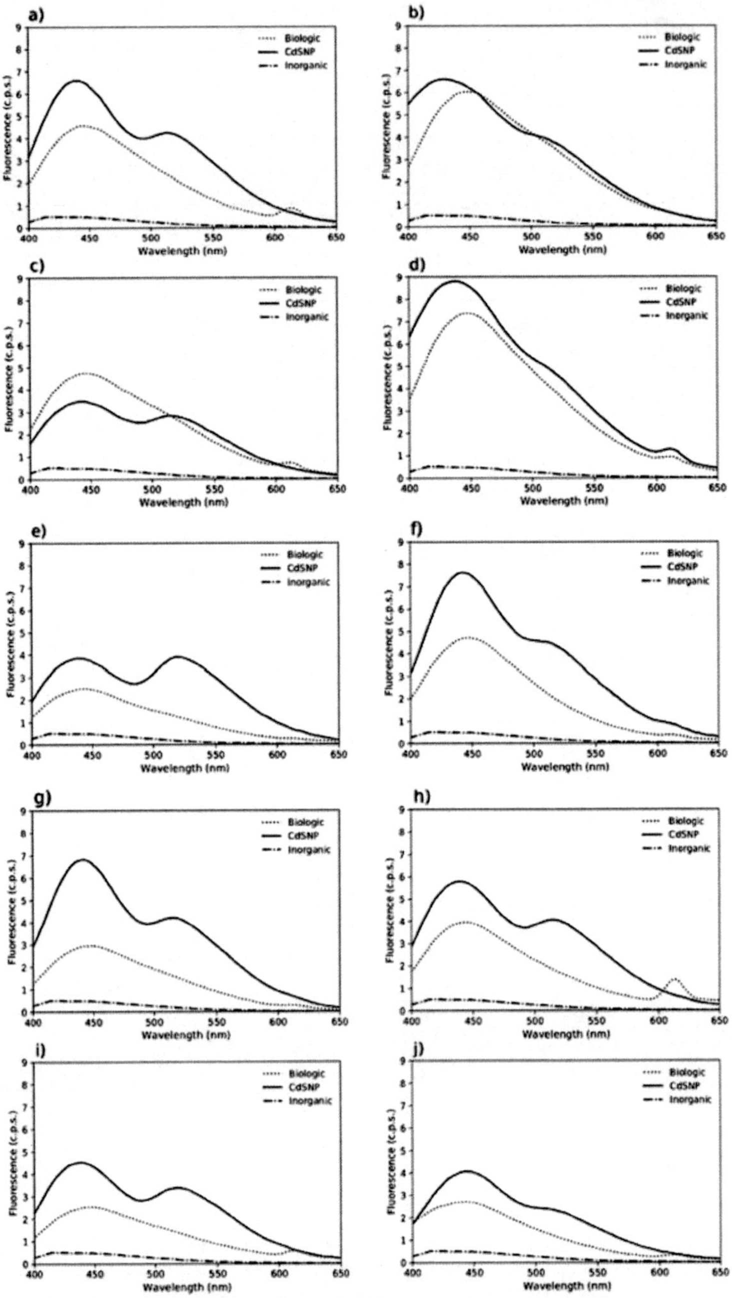

Fig. 3 Fluorescence spectrophotometry analysis. Fungal isolates are indicated in different graphs: **a** ZML1; **b** ZML2; **c** ZML3; **d** ZML5; **e** MML3; **f** MMS2; **g** CAS3.2; **h** CAS3; **i** CAL2; and **j** CAL3. Negative biologic and inorganic controls are coded as "Biologic" and "Inorganic", whereas the sample where CdSNP synthesis reaction was done is indicated as "CdSNP"

Table 2 Dynamic light scattering analysis for isolate samples

Isolate samples	Hydrodynamic radius (nm)	Polydispersity index (%)	pH
ZML1	1040.74 ± 422.98	36.17 ± 6.35	6.73
ZML5	564.67 ± 312.56	40.86 ± 4.87	8.32
MMS2	383.71 ± 28.30	22.96 ± 3.60	8.16
CAS3.2	2383.52 ± 1601.90	31.81 ± 4.44	7.62

300 to 2000 nm, as well as a polydispersity index > than 20%, corresponding to organic material remains that pass through the 0.20 μm filters [26]. Samples of MMS2 and ZML5 isolates presented an average hydrodynamic radius around 300 and 500 nm approximately; MMS2 isolate showed the lowest percentage of polydispersity (Table 2). It is important to mention that previous results may be influenced by the pH in the solution, where greater colloidal stability of the particles could be generated around a pH value of 8 [26].

Frequency distribution analysis of the particles present in the fungal samples, showed the highest relative particle frequency and the largest error in the range of 500–1000 nm for hydrodynamic radius, except for MMS2 sample, which had the highest frequency in the range of 200–500 nm, indicating that particles greater than 500 nm are not uniform or constant, since they could have traces of organic material and conglomerated particles. Only in the MMS2 and ZML5 isolates, particles between 100 and 200 nm of hydrodynamic radius were detected, while a low frequency of particles with hydrodynamic radius below 100 nm in four fungal isolates was found. These correspond to 3.69 ± 1.51, 11.16 ± 4.06, 13.41 ± 8.15 and 20.69 ± 14.64% of cumulative frequency for ZML1, ZML5, MMS2 and CAS3.2 isolates, respectively (Fig. 4). In the case of functionalized nanoparticles, the hydrodynamic radius determines the total diameter of the nanoparticle, including the core and the shell [39]. Therefore, when particles are capped with various biological components, their size can increase considerably. However, the actual size of the nanoparticle core could only be confirmed by electron microscopy [26].

4 Conclusion

The CdSNP synthesis by cadmium-tolerant fungi was evidenced by UV-Vis and fluorescence spectrophotometry. Dynamic light scattering demonstrated the obtention of particles less than 100 nm for isolates ZML1, ZML5, MMS2 and CAS3.2.

The analysis of hydrodynamic radius distribution of particles demonstrated the presence of considerable amounts of organic debris, as well as particles with a hydrodynamic radius less than 100 nm for four fungal isolates. According with results, cadmium-tolerant fungi have cellular mechanisms that allow them to grow in high cadmium concentrations then facilitating the synthesis of nanometric particles, although additional analyses are needed to confirm their size and core. For previous

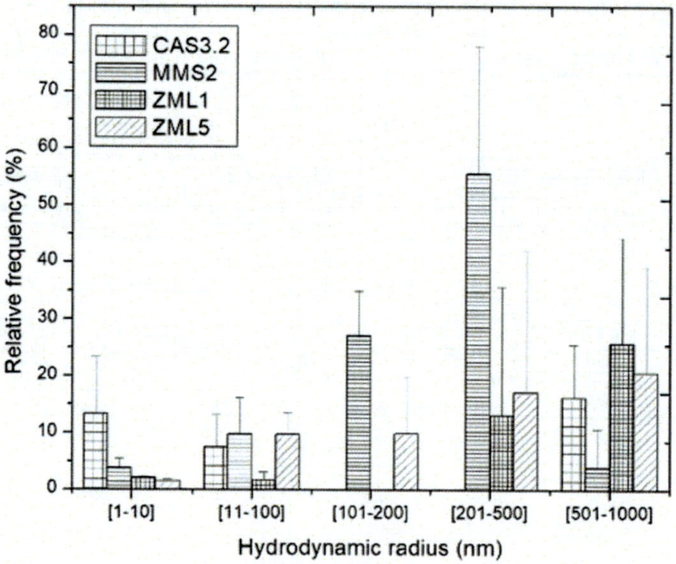

Fig. 4 Hydrodynamic radius particles distribution

reason, cadmium-tolerant fungi could be considered as promising organisms for metallic nanoparticles synthesis, particularly cadmium sulfide ones.

Acknowledgements The research was financially by Grant No. A1-S-31777 from Consejo Nacional de Ciencia y Tecnología and Instituto Politécnico Nacional. Authors acknowledge to Anton Paar Mexico for support and technical assistant for the achievement of the present study.

References

1. Kumar M (2019) Semiconductor nanoparticles theory and applications. Int J Appl Eng Res 14:491–494
2. Suresh S (2013) Semiconductor nanomaterials, methods and applications: a review. Nanosci Nanotechnol 3:62–74. https://doi.org/10.5923/j.nn.20130303.06
3. Pawar RS, Upadhaya PG, Patravale VB (2018) Quantum dots: novel realm in biomedical and pharmaceutical industry. In: Hussain CM (ed) Handbook of nanomaterials for industrial application. Micro and nano technologies, vol 2018. Elsevier Inc., Newark, pp 621–637. https://doi.org/10.1016/B978-0-12-813351-4.00035-3
4. Meziani MJ, Pathak P, Harruff BA, Hurezeanu R, Sun Y, Carolina S (2005) Direct conjugation of semiconductor nanoparticles with proteins. Langmuir 21:2008–2011
5. Yuting L, Jing Z, Donghui L (2019) Preparation of cadmium sulfide nanoparticles and their application for improving the properties of the electrochemical sensor for the determination of enrofloxacin in real samples. Chirality 31:174–184. https://doi.org/10.1002/chir.23045
6. Bera D, Qian L, Tseng TK, Holloway PH (2010) Quantum dots and their multimodal applications: a review. Materials (Basel) 3:2260–2345

7. Sandoval-Cárdenas I, Gómez-Ramírez M, Rojas-Avelizapa NG, Vidales-Hurtado MA (2017) Synthesis of cadmium sulfide nanoparticles by biomass of *Fusarium oxysporum f. sp. lycopersici.* J Nano Res 46:179–191. https://doi.org/10.4028/www.scientific.net/JNanoR.46.179
8. Sharma G, Pandey S, Ghatak S, Watal G, Rai PK (2018) Potential of spectroscopic techniques in the characterization of "green nanomaterials". In: Durgesh KT, Parvaiz A, Shivesh S, Devendra KC, Nawal KD (eds) Nanomaterials in plants, algae, and microorganisms. Concepts and controversies, vol 1. Elsevier Inc., pp 59–77. https://doi.org/10.1016/B978-0-12-811487-2.00003-7
9. Hulkoti NI, Taranath TC (2014) Biosynthesis of nanoparticles using microbes—a review. Colloids Surf B Biointerfaces 121:474–483
10. Ahmad A, Mukherjee P, Mandal D, Senapati S, Khan MI, Kumar R, Sastry M (2002) Enzyme mediated extracellular synthesis of CdS nanoparticles by the fungus, *Fusarium oxysporum.* J Am Chem Soc 124:12108–12109. https://doi.org/10.1021/ja027296o
11. Siddiqi KS, Husen A (2016) Fabrication of metal nanoparticles from fungi and metal salts: scope and application. Nanoscale Res Lett 11:1–15. https://doi.org/10.1186/s11671-016-1311-2
12. Mal J, Nancharaiah YV, Van Hullebusch ED, Lens PNL (2016) Metal chalcogenide quantum dots: biotechnological synthesis and applications. RSC Adv 6:41477–41495. https://doi.org/10.1039/c6ra08447h
13. Durán N, Marcato PD, Alves OL, De Souza GIH, Esposito E (2005) Mechanistic aspects of biosynthesis of silver nanoparticles by several *Fusarium oxysporum* strains. J Nanobiotechnol 3:1–7. https://doi.org/10.1186/1477-3155-3-8
14. Reyes LR, Gómez I, Garza MT (2009) Biosynthesis of cadmium sulfide nanoparticles by the Fungi *Fusarium sp.* Int J Green Nanotechnol Biomed 1:90–95. https://doi.org/10.1080/19430850903149936
15. Jacob JM, Lens PNL, Balakrishnan RM (2016) Microbial synthesis of chalcogenide semiconductor nanoparticles: a review. Microb Biotechnol 9:11–21. https://doi.org/10.1111/1751-7915.12297
16. Dameron CT, Reese RN, Mehra RK, Kortan AR, Carroll PJ, Steigerwald ML, Brus LE, Winge DR (1989) Biosynthesis of cadmium sulphide quantum semiconductor crystallites. Nature 338:596–597. https://doi.org/10.1038/338596a0
17. Dameron C, Winge D (1990) Characterization of peptide-coated cadmium-sulfide crystallites. Inorg Chem 29:1343–1348. https://doi.org/10.1021/ic00332a011
18. Borovaya M, Pirko Y, Krupodorova T, Naumenko A, Blume Y, Yemets A (2015) Biosynthesis of cadmium sulphide quantum dots by using *Pleurotus ostreatus* (Jacq.) P. Kumm. Biotechnol Biotechnol Equip 29:1156–1163. https://doi.org/10.1080/13102818.2015.1064264
19. Chen G, Yi B, Zeng G, Niu Q, Yan M, Chen A, Du J, Huang J, Zhang Q (2014) Facile green extracellular biosynthesis of CdS quantum dots by white rot fungus *Phanerochaete chrysosporium.* Colloids Surf B Biointerfaces 117:199–205. https://doi.org/10.1016/j.colsurfb.2014.02.027
20. Sanghi R, Verma P (2009) A facile green extracellular biosynthesis of CdS nanoparticles by immobilized fungus. Chem Eng J 155:886–891. https://doi.org/10.1016/j.cej.2009.08.006
21. Bhadwal AS, Tripathi RM, Gupta RK, Kumar N, Singh RP, Shrivastav A (2014) Biogenic synthesis and photocatalytic activity of CdS nanoparticles. RSC Adv 4:9484–9490. https://doi.org/10.1039/c3ra46221h
22. Das SK, Shome I, Guha AK (2012) Surface functionalization of *Aspergillus versicolor* mycelia: *in situ* fabrication of cadmium sulphide nanoparticles and removal of cadmium ions from aqueous solution. RSC Adv 2:3000–3007. https://doi.org/10.1039/c2ra01273a
23. Uddandarao P, Mohan R (2016) ZnS semiconductor quantum dots production by an endophytic fungus *Aspergillus flavus.* Mater Sci Eng B Solid-State Mater Adv Technol 207:26–32. https://doi.org/10.1016/j.mseb.2016.01.013
24. Sandoval-Cárdenas I, Gómez-Ramírez M, Rojas-Avelizapa NG (2017) Use of a sulfur waste for biosynthesis of cadmium sulfide quantum dots with *Fusarium oxysporum f. sp. lycopersici.* Mater Sci Semicond Process 63:33–39. https://doi.org/10.1016/j.mssp.2017.01.017

25. Inouhe M, Sumiyoshi M, Tohoyama H, Joho M (1996) Resistance to cadmium ions and formation of a cadmium-binding complex in various wild-type yeasts. Plant Cell Physiol 37:341–346. https://doi.org/10.1093/oxfordjournals.pcp.a028951
26. Khandel P, Shahi SK (2018) Mycogenic nanoparticles and their bio-prospective applications: current status and future challenges. J Nanostruct Chem 8:369–391. https://doi.org/10.1007/s40097-018-0285-2
27. Moghaddam AB, Namvar F, Moniri M, Tahir PM, Azizi S, Mohamad R (2015) Nanoparticles biosynthesized by fungi and yeast: a review of their preparation, properties, and medical applications. Molecules 20:16540–16565
28. Kelly RT, Love NG (2007) Ultraviolet spectrophotometric determination of nitrate: detecting nitrification rates and inhibition. Water Environ Res 79:808–812. https://doi.org/10.2175/106143007x156682
29. Raynal B, Lenormand P, Baron B, Hoos S, England P (2010) Quality assessment and optimization of purified protein samples: why and how? Microb Cell Fact 13:180. https://doi.org/10.1186/s12934-014-0180-6
30. Taniguchi M, Lindsey JS (2018) Database of absorption and fluorescence spectra of >300 common compounds for use in photochem CAD. Photochem Photobiol 94:290–327. https://doi.org/10.1111/php.12860
31. Bhatia S, Garg A, Sharma K, Kumar S, Sharma A, Purohit AP (2011) Mycosporine and mycosporine-like amino acids: a paramount tool against ultra violet irradiation. Pharmacogn Rev 5:138–146
32. Lawaetz AJ, Stedmon CA (2009) Fluorescence intensity calibration using the Raman scatter peak of water. Appl Spectrosc 63:936–940. https://doi.org/10.1366/000370209788964548
33. Perucho J, Gonzalo-Gobernado R, Bazan E, Casarejos MJ, Jiménez-Escrig A, Asensio MJ, Herranz AS (2015) Optimal excitation and emission wavelengths to analyze amino acids and optimize neurotransmitters quantification using precolumn OPA-derivatization by HPLC. Amino Acids 47:963–973. https://doi.org/10.1007/s00726-015-1925-1
34. Alex S, Le Thanh H, Vocelle D (1992) Studies of the effect of hydrogen bonding on the absorption and fluorescence spectra of all-trans-retinal at room temperature. Can J Chem 70:880–887. https://doi.org/10.1139/v92-117
35. Gadalla A, Almokhtar M, Abouelkhir AN (2018) Effect of Mn doping on structural, optical and magnetic properties of CdS diluted magnetic semiconductor nanoparticles. Chalcogenide Lett 15:207–218
36. Su J, Zhang T, Li Y, Chen Y, Liu M (2016) Photocatalytic activities of copper doped cadmium sulfide microspheres prepared by a facile ultrasonic spray-pyrolysis method. Molecules 21:735. https://doi.org/10.3390/molecules21060735
37. Pan YL (2015) Detection and characterization of biological and other organic-carbon aerosol particles in atmosphere using fluorescence. J Quant Spectrosc Radiat Transf 150:12–35. https://doi.org/10.1016/j.jqsrt.2014.06.007
38. Kalra R, Conlan XA, Goel M (2020) Fungi as a potential source of pigments: harnessing filamentous fungi. Front Chem 8:369
39. Zheng T, Bott S, Huo Q (2016) Techniques for accurate sizing of gold nanoparticles using dynamic light scattering with particular application to chemical and biological sensing based on aggregate formation. ACS Appl Mater Interfaces 8:21585–21594. https://doi.org/10.1021/acsami.6b06903

Reducing Power of *Curcuma longa* Extract and Its Influence on the Synthesis of Copper Nanoparticles

I. A. Cruz-Rodríguez, A. M. Rivas-Castillo, and N. G. Rojas-Avelizapa

Abstract Nowadays, green synthesis has gained ground over chemical or physical methods to synthesize nanoparticles, not only because of its advantages, such as lower costs and environmental friendliness, but also because of the possibility to prevent oxidation and provide stability all in the same process. However, only generalities are known about plant extracts as reducing agents. Therefore, more studies to understand reducing power during green synthesis are needed. Thus, this study evaluated different concentrations of copper as a precursor material for the synthesis of nanoparticles and its relationship with the reducing power present in *Curcuma longa* extract, used during this synthesis. Results showed that the reducing power of *Curcuma longa* extract is being used during the synthesis of copper nanoparticles and that the concentrations of precursor material (copper) affected the visual appearance of final solutions which could impact in other nanoparticles characteristics.

Keywords CuNPs · Curcuma · FRAP · Green synthesis

1 Introduction

Metallic nanoparticles (NPs) have been widely studied. Among them, copper nanoparticles (CuNPs), whether in their metallic or oxidized forms, are of great interest because of the broadness of their applications; some examples are conductive inks, batteries, solar cells, sensing (bio-sensing, gas sensing, electrochemical sensing), water treatment, as bactericidal and antifungal agents, and as anticancer agents as well [1].

Synthesis of CuNPs can be done through different methods, being the most common the polyol process, metal vapor synthesis, laser ablation, micro-emulsion

I. A. Cruz-Rodríguez · N. G. Rojas-Avelizapa (✉)
CICATA-QRO-IPN, Querétaro, México
e-mail: nrojasa@ipn.mx

I. A. Cruz-Rodríguez · A. M. Rivas-Castillo · N. G. Rojas-Avelizapa
Universidad Tecnológica de la Zona Metropolitana del Valle de México (UTVAM), Tizayuca, México

K. L. Flores Rodríguez et al. (eds.), *Recent Trends in Sustainable Engineering*, Lecture Notes in Networks and Systems 297,
https://doi.org/10.1007/978-3-030-82064-0_13

techniques, chemical reduction, and thermal reduction. Nevertheless, these methods are both expensive and hazardous for the environment [1]. Therefore, less harmful alternatives have been studied such as green synthesis, which includes the use of microorganisms and plants [2]. Specifically, green synthesis using plants is more popular than the use of microorganisms, since the process may be simpler and costs are lower; also, synthesis may be carried out faster and energy needed is lower. Besides, biomolecules acting as capping agents for NPs could stabilize and prevent oxidation, and diminish agglomeration of NPs as well [3].

Because of the complexity of the metabolites involved, it cannot be certainly known which specific metabolites are responsible for nanoparticle synthesis. However, in general terms, flavonoids, phenols, and sterols (all reducing agents) are considered responsible for redox reactions which have an impact on nanoparticle formation. As for the production of CuNPs, some plants or parts of them like green and black tea (*Camellia sinensis*), leafy spurge (*Euphorbia esula* L.), pistachio hull (*Pistacia vera* L.), magnolia leaf (*Magnolia* spp.), pomegranate (*Punica granatum)*, chinese hibiscus (*Hibicus rosa-sinensis*), false daysi (*Eclipta prostrata*), ceylon caper (*Capparis zeylanica*), flame lily (*Gloriosa superba*), moringa (*Moringa oleifera*) and curcuma (*Curcuma longa*) have been studied [1, 3, 4], and there have been observed differences in nanoparticle characteristics among all evaluated plants, which could be related to reducing power of plant extracts, the metabolites involved in the process, or the synthesis conditions used.

Thus, the present work used an environmentally friendly methodology for the synthesis of CuNPs using copper acetate dihydrate as precursor material and *Curcuma longa* extract as both reducing and capping agent.

2 Materials and Methods

During this experimentation, different concentrations of precursor material (copper) were evaluated while keeping the *Curcuma longa* extract constant. In order to do so, 50 mL of ($Cu(CH_3COO)_2 \cdot H_2O$) 0.1 M were prepared as stock solution for the synthesis reproduction. Later different concentrations of precursor material were prepared (100, 200, 300, 400 and 1000 mg/L of copper). Besides, ferric reducing antioxidant power assay (FRAP) was performed to know the reducing power present in the *Curcuma longa* extract used for the synthesis, and the supernatants collected after the synthesis of CuNPs.

2.1 *Preparation of the* Curcuma Longa *Extract*

Curcuma longa tubers were washed five times with tap water to eliminate mud and other residues. Then tubers were washed twice again using distilled water. Afterward, they were dried for over half a week in the oven at 70 °C, to eliminate most of

Table 1 Experimental conditions for the synthesis of CuNPs

Sample No.	Curcuma extract (mL)	Concentration of copper (mg/L)
1	25	1000
2	25	400
3	25	300
4	25	200
5	25	100

the moisture and then pulverized by blending, and they were, once more, dried at room temperature. Once dried, *Curcuma longa* extract was prepared following the methodology previously reported [5]. Briefly, 10 g of *Curcuma longa* powder were added to 100 mL of ethanol in a 250 mL flask; the solution was stirred for 4 h at 70 °C on a hot plate. Finally, *Curcuma longa* extract was filtered using fine pore filter paper (1 μm).

2.2 Green Synthesis of CuNPs

For green synthesis, a modification of the methodology implemented by Jayarambabu et al. [5] was followed, where 50 mL of copper acetate dihydrate solution was used as precursor material and 25 mL of *Curcuma longa* extract as reducing agent (extract was used in a period no longer than 7 days to preserve properties). Different concentrations of the precursor material were evaluated (100, 200, 300, 400, and 1000 mg/L of copper) while keeping constant the use of 25 mL of Curcuma extract in each synthesis. Conditions of the synthesis are shown in Table 1. Both the precursor material and the extract were mixed and stirred for 15 min on a hot plate at room temperature. After stirring time, the solution was put into the microwave oven (SAMSUNG MR123C) for 180 s at the 2-potential level, to reach around 200 W power (like the methodology followed). Then, the five syntheses samples were centrifuged for 10 min at 6000 rpm to separate supernatants from the pellets (presumably CuNPs). Subsequently, pellets were washed with ethanol once. Finally, they were collected and put into the oven for 24 h at 70 °C for drying.

2.3 UV–Vis Characterization

To corroborate the synthesis of CuNPs through surface plasmon resonance (SPR), 1 mg of pellet from each synthesis was re-suspended in ethanol. Subsequently, UV–Vis analyses were performed by measuring from 400 to 700 nm in a UV–Vis spectrophotometer (GENESYS 10S, Thermo Scientific).

2.4 Ferric Reducing Antioxidant Power (FRAP)

Ferric reducing antioxidant power assay (FRAP) was performed in the *Curcuma longa* extract and the supernatants obtained after syntheses to assess differences in the amount of their reducing power. Solutions were prepared as indicated by Vijayalaksshmi and Ruckmani [6]. To this end, 1 mL of each sample was vortexed with 2.5 mL of phosphate buffer (pH 6.6) and 2.5 mL of potassium ferrocyanide. Then, solutions were incubated at 50 °C for 20 min. After incubation, 2.5 mL of trichloroacetic acid was added, and the solutions were centrifuged at 6000 rpm for 10 min. After-ward, 2.5 mL of supernatant, 2.5 mL of water, and 0.5 mL of ferric chloride solution were mixed thoroughly. Finally, absorbance was immediately measured after incubation of 10 min at 700 nm using a UV–Vis spectrophotometer (GENESYS 10S, Thermo Scientific).

2.5 Statistical Analysis

Basic statistical parameters and analysis of variance (one-way ANOVA) was performed using the commercial statistical software OriginPro 9.0. Differences in P values ≤ 0.05 were considered as statistically significant, and lower-case letters represent groups of statistically different data.

3 Results and Discussion

3.1 Synthesis of CuNPs

To corroborate the formation of CuNPs, the color change of the solution was observed during the first 10 min, followed by the appearance of a new peak at 447 nm through UV–Vis analyses, and the results obtained were compared to the ones previously reported by the methodology followed, where 50 mL of *Curcuma longa* extract were put in contact with copper acetate dihydrate solution (0.1 M/100 ml) [5]. Change in color from yellow to brown was observed during synthesis, in accordance with previous results. Then, UV–Vis analyses were performed from 400 to 700 nm on the *Curcuma longa* extract (Fig. 1) where is observed a peak around 426 nm, and the precursor material (Fig. 2), where no peak was found. Later, the UV–Vis analyses were done on the re-suspended pellet after synthesis (Fig. 3), in order to find differences on the already existing peak or appearance of new ones. In Fig. 3, a new peak can be observed around 447 nm. However, the appearance of this peak does not correspond to previous observations, where a peak appears at 524 nm, or other reported peaks at 270 and 440 nm [7], or 260 nm [8]. Additionally, it was reported that CuNPs around 10–40 nm of diameter usually show their SPR at 560 nm, and

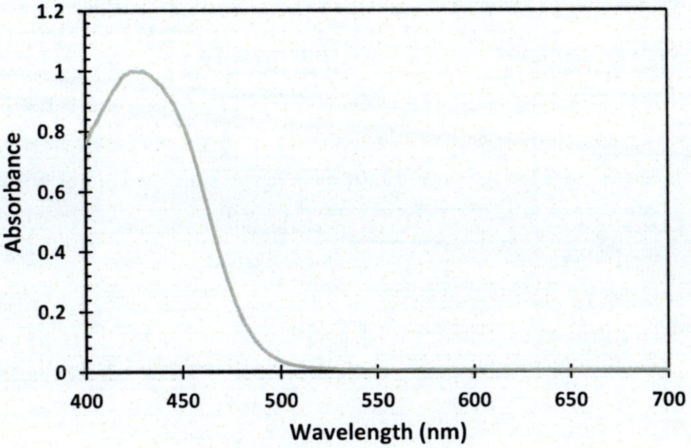

Fig. 1 UV–Vis spectrum of *Curcuma longa* extract

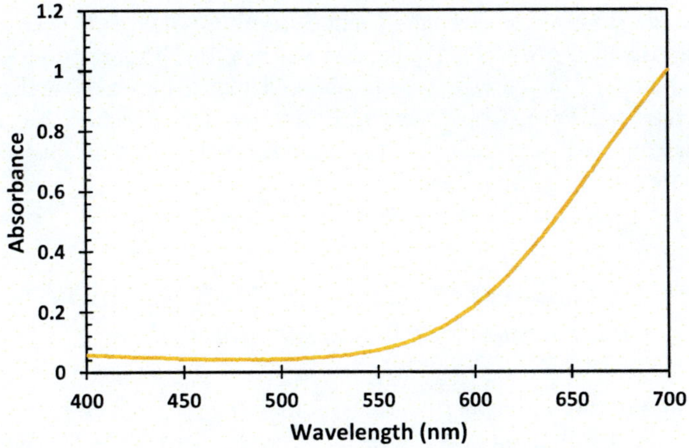

Fig. 2 UV–Vis spectrum of precursor material (copper acetate dihydrate solution)

in some cases, NPs are smaller than 20 nm, which cannot be detected by SPR [9]. Therefore, there can be assumed that there is not a unique or specific peak for CuNPs formation. Besides, the difference in appearance between the band detected in this study and the cited previous reports could be related to differences in some steps of the procedure, like plant obtention (due to changes in metabolites from region to region), drying process of tubers (that could have altered metabolites), or power used during microwave oven contact (which impacts in the amount of the temperature reached).

Hence, the appearance of a new peak at 447 nm, coupled with the change in color after the synthesis process, could suggest the formation of CuNPs. However, more

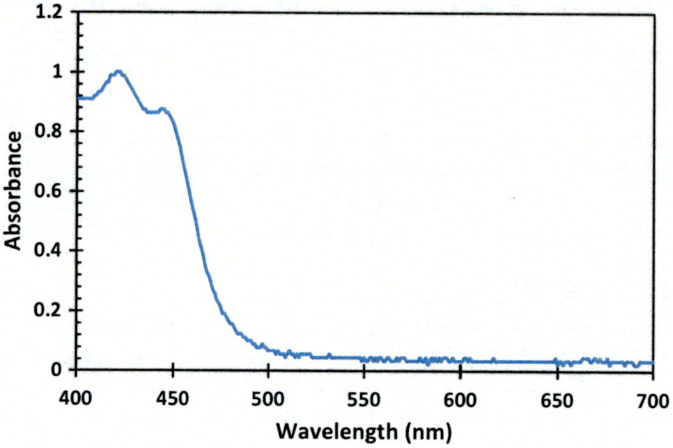

Fig. 3 UV–Vis spectrum of CuNPs

specific studies like Dynamic Light Scattering (DLS) and Transmission Electron Microscopy (TEM) should be necessary to confirm NPs formation. Replication of the methodology for the synthesis of CuNPs enabled to have a reference of the behaviors (coloration, UV–Vis) observed. Once this reference was established, experimentation changing the concentrations of the precursor material was evaluated, as is described below.

3.2 *Evaluation of CuNPs Formation by Changing Concentrations of Precursor Material*

Solutions varying copper concentrations were prepared from a stock solution to evaluate the minimum concentration of copper for CuNPs formation, and observe possible changes (coloration, SPR) after synthesis. It is known that some indicators of NPs formation can be the change in color of the solutions after synthesis and the appearance of new peak formations that can be seen through UV–Vis analyses, as shown in Fig. 3. In the case of CuNPs, most authors report sample darkening at the end of synthesis [3, 9–12] while others have reported yellow-ocher colors after synthesis procedures [1, 13]. In this study, three types of coloration were observed after the synthesis process, which were dark brown, light brown, and yellow, as shown in Fig. 4. The dark brown solutions had copper concentrations above 300 mg/L, light brown was observed with 200 mg/L, and yellow with 100 mg/L of copper. Syntheses done with 100 and 200 mg/L of copper neither reached dark brown coloration nor precipitated. Thus, from these observations, it can be suggested that copper concentrations under 200 mg/L may not be enough for a complete synthesis of nanoparticles, and that other properties like morphology and size could also be different.

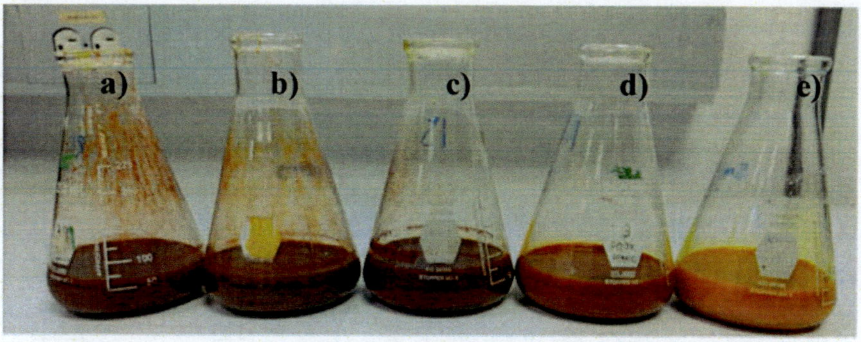

Fig. 4 Change in color after synthesis of CuNPs according to the concentrations evaluated: **a** 1000 mg/L, **b** 400 mg/L, **c** 300 mg/L, **d** 200 mg/L and **e** 100 mg/L of copper

3.3 UV–Vis Analyses

To confirm surface plasmon resonance of CuNPs, UV–Vis analyses were performed, evidencing two behaviors. First, as can be observed in Fig. 5, the absorption spectrum shows that the higher copper concentration, the lower absorbance, and vice versa. It is suggested that high absorbances in the spectra after synthesis could be attributed to an excess of antioxidant power still present on the supernatant, which was not used during copper particle reduction. Second, as shown in Fig. 6, where there are presented the UV–Vis analyses of the lowest and the highest concentration of copper acetate evaluated, it can be observed that differences in the copper concentration used for NPs synthesis impact the peak definition at 447 nm. It may be possible that synthesis using low copper concentrations (100 and 200 mg/L) as precursor material,

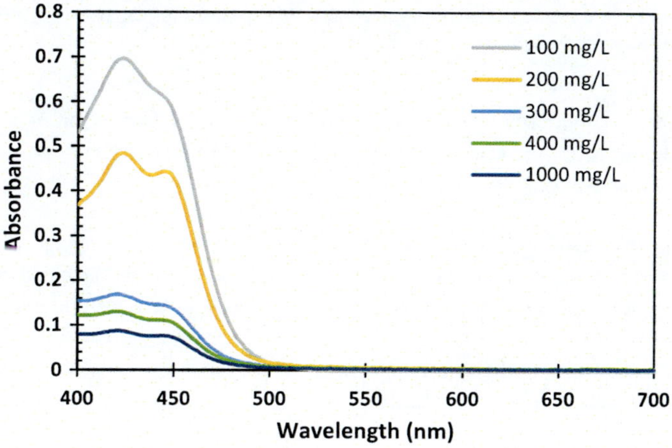

Fig. 5 UV–Vis spectra of CuNPs according to the copper concentrations evaluated

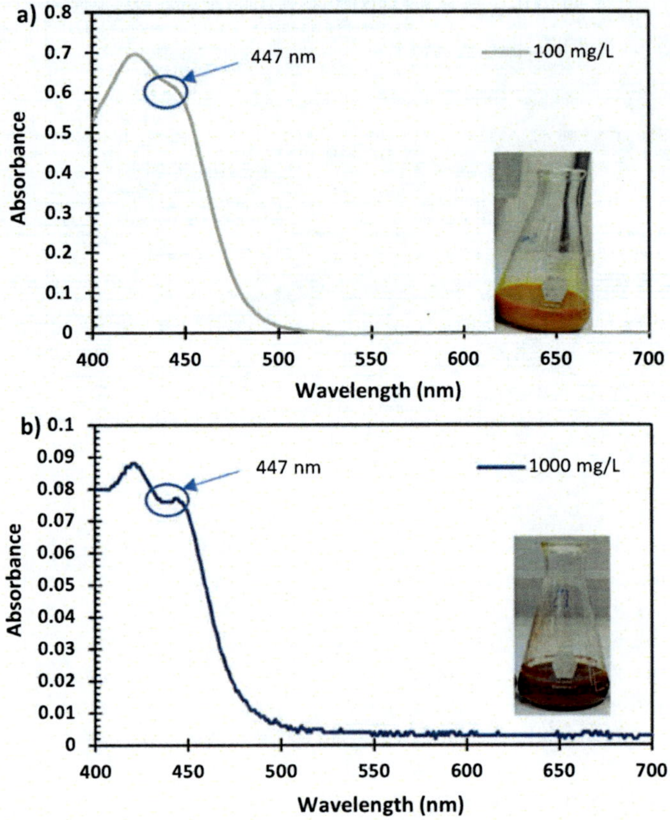

Fig. 6 Comparison between CuNPs synthesis at the lowest (100 mg/L) and highest (1000 mg/L) copper concentrations evaluated. **a** UV–Vis spectrum of CuNPs with 100 mg/L of copper, additional photo shows solution after synthesis; **b** UV–Vis spectrum of CuNPs with 1000 mg/L, additional photo shows solution after synthesis

where peaks at 447 nm are not so defined, allow the formation of NPs, but that these concentrations are not enough to cause a dark brown color change, like clearly shown in concentrations above 300 mg/L.

Ferric Reducing Antioxidant Power (FRAP)

The synthesis of various metallic NPs is already possible by chemical and physical methods. However, green synthesis has gained ground over them, and the main reasons are the cost reduction they represent and the fact of being considered environmental-friendly. Nevertheless, little is known about how syntheses using plant extracts or microorganisms work, mainly due to the number of metabolites involved during these kinds of processes. Some authors refer to the reducing power of plant extracts as the responsible for NPs formation. Therefore, theoretically, any kind of reducing agent should do the work, but this does not seem to apply in all cases. It is reported that various plant extracts, like the ones from Indian jujube (*Ziziphus mauritiana*), garlic (*Allium sativum*), neem (*Azadirachta indica*), mango (*Mangifera indica*), banyan tree (*Ficus benghalensis*), papaya (*Carica papaya*), and orange (*Citrus sinensis*), show negative results for the synthesis of gold NPs [14]. However, other authors differ like in the case of *Azadirachta indica* extract, which has been useful to synthesize silver NPs [15]; it has also been reported the formation of bimetallic nanoparticles (Au–Ag) with *Azadirachta indica* extract [16]. As was mentioned before, the biomolecules related to the reducing power properties of plant extracts may have an impact on size reduction, be used as capping agents, and could also affect NPs morphology. Specifically, there has been evaluated the reducing power from devil's backbone (*Kalanchoe daigremontiana*) extracts [4], which were elaborated with diverse solvents, such as distilled water, ethanol, and isopropanol. These extracts showed different amounts of reducing power as well as the liberation of different metabolites, according to the polarity of these solvents. Those differences seemed to impact both the size and morphology of the NPs synthesized. Therefore, more studies should focus on assessing and determining the reducing power of plant extracts, to understand better the very complex stoichiometric reactions involved, how this could affect NPs characteristics, and if this ability to reduce to NPs differs depending on the evaluated metal (precursor material).

Additionally, to evaluate if there was a change in the reducing power of the synthetized solution after the synthesis of CuNPs, a FRAP assay was performed on the Curcuma extract (reducing agent) and the five supernatants, corresponding to the synthesis of CuNPs, where different copper concentrations were evaluated. Ascorbic acid standards were prepared to calculate the concentration of antioxidants in samples (reducing power). Figure 7 shows the results obtained, and there can be observed yellow to green color changes, depending on the concentration of antioxidants present in the samples, where darker colors indicate a higher presence of antioxidants. Reducing power of *Curcuma longa* extract was found to be equivalent to 3200 mg/L of ascorbic acid during this study. This is considered as the initial amount of reducing power involved during CuNPs synthesis. Additionally, FRAP assay was performed on the supernatants obtained after syntheses, and results are shown in Fig. 8, observing that, in the cases where the supernatants had copper concentrations over 400 mg/L, 100% of reducing power (expressed in mg/mL of ascorbic acid) from the extract was consumed. On the contrary, in syntheses where copper concentrations were less than 300 mg/L, there was still reducing power. Specifically, 300 mg/L consumed 99.77% of reducing power while in supernatants with 100 and 200 mg/L of copper, 94.82 and 94.74% of reducing power was consumed, respectively. Correlation data from the 5 supernatants analyzed was −0.66, indicating an inverse relationship between reducing power and copper concentration in synthesis. In this regard, a similar behavior (diminishing of reducing power) has been reported before [8]; this previous report studied the amount of reducing power after NPs synthesis using moringa leaves (*Moringa oleifera*), finding that there was a reduction in total polyphenol concentration, which was equivalent to reducing power, deducing that some polyphenols were involved in the bio-reduction of NPs. Therefore, these results strongly suggest the use of reducing power of *Curcuma longa* extract during the synthesis of CuNPs and a possible stoichiometric reaction between the amount of precursor material and the reducing power consumed during this process.

Due to the changes in color of the solutions (brownish, ocher), and the appearance of a new peak formation at 447 nm, it is suggested that green synthesis of CuNPs using the ethanolic extract of *Curcuma longa* tubers was successful. FRAP assay, done on the supernatants with different concentrations of precursor material, indicate that in the cases where concentrations were higher, all the reducing power present in the extracts was consumed, while lower concentrations of precursor material still had

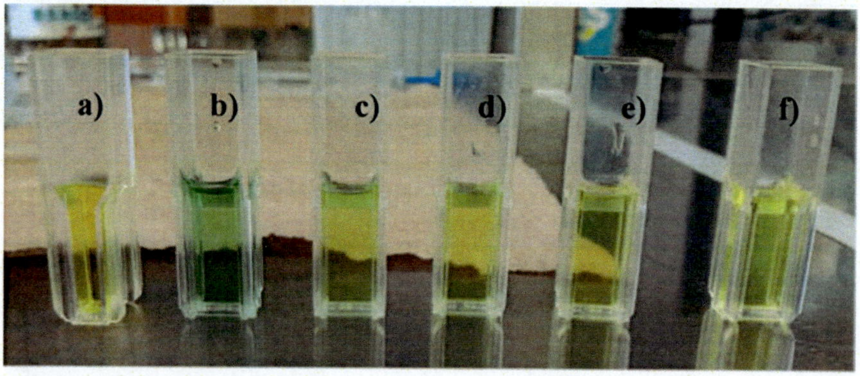

Fig. 7 FRAP assay of CuNPs supernatants with different copper concentrations and *Curcuma longa* extract. **a** 1000 mg/L, **c** 100 mg/L, **d** 200 mg/L, **e** 300 mg/L, **f** 400 mg/L of copper and **b** *Curcuma longa* extract

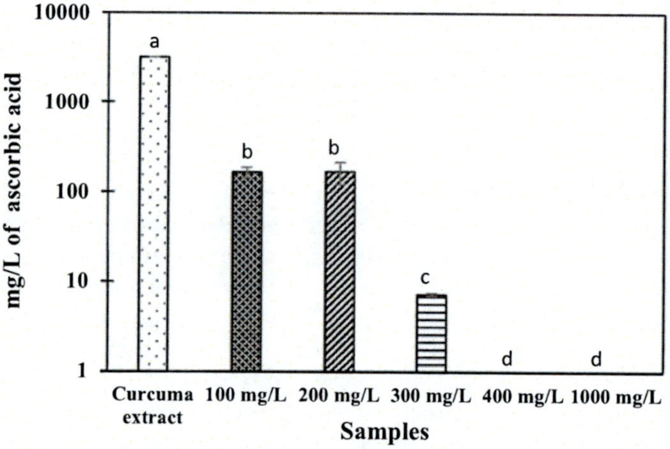

Fig. 8 Total reducing power found in supernatants after CuNPs synthesis (varying copper concentrations) and *Curcuma longa* extract expressed in mg/mL of ascorbic acid. Reducing power was not detected in supernatants with copper acetate concentrations of 400 and 1000 mg/L. Letters indicate significative differences between samples

reducing power left in the extracts. Therefore, the results obtained strongly suggest that the reducing power of *Curcuma longa* extract is involved during the synthesis of CuNPs.

Further studies such as DLS and TEM would be required to complement and confirm the formation of CuNPs, as well as to study possible differences in morphology between the concentrations evaluated.

4 Conclusions

Color changes in all synthetized solutions (brownish, ocher) and the appearance of a new peak at 447 nm, suggest that CuNPs can be successfully synthetized with ethanolic extract of *Curcuma longa*. Differences in the amount of reducing power found on the supernatants at the end of experimentation using different copper concentrations for CuNPs synthesis may indicate that: (a) reducing power is responsible of NPs formation, and (b) there is a correlation between the amount of reducing power and the precursor material used during synthesis.

Acknowledgements The authors acknowledge the support of CONACYT scholarship, PIFI project 20201305, and the Research Center in Applied Science and Advanced Technology (CICATA-QRO-IPN).

References

1. Sampaio S, Viana JC (2021) Optimisation of the green synthesis of Cu/Cu_2O particles for maximum yield production and reduced oxidation for electronic applications. Mater Sci Eng B 263(2020):114807. https://doi.org/10.1016/j.mseb.2020.114807
2. Rana A, Yadav K, Jagadevan S (2020) A comprehensive review on green synthesis of nature-inspired metal nanoparticles: mechanism, application and toxicity. J Clean Prod 272:122880. https://doi.org/10.1016/j.jclepro.2020.122880
3. Wang G et al (2021) Green synthesis of copper nanoparticles using green coffee bean and their applications for efficient reduction of organic dyes. J Environ Chem Eng 9(4):105331. https://doi.org/10.1016/j.jece.2021.105331
4. Vergara-Castaneda H et al (2019) Gold nanoparticles bioreduced by natural extracts of arantho (Kalanchoe daigremontiana) for biological purposes: physicochemical, antioxidant and antiproliferative evaluations. Mater Res Express 6(5):55010. https://doi.org/10.1088/2053-1591/ab0155
5. Jayarambabu N, Akshaykranth A, Rao TV, Rao KV, Kumar RR (2020) Green synthesis of Cu nanoparticles using Curcuma longa extract and their application in antimicrobial activity. Mater Lett 259:126813. https://doi.org/10.1016/j.matlet.2019.126813
6. Vijayalakshmi M, Ruckmani K (2016) Ferric reducing anti-oxidant power assay in plant extract. Bangladesh J Pharmacol 11(3):570–572. https://doi.org/10.3329/bjp.v11i3.27663
7. Rashad M, Rüsing M, Berth G, Lischka K, Pawlis A (2013) CuO and Co_3O_4 nanoparticles: synthesis, characterizations, and raman spectroscopy. J Nanomater 2013. https://doi.org/10.1155/2013/714853

8. Das PE, Abu-Yousef IA, Majdalawieh AF, Narasimhan S, Poltronieri P (2020) Green synthesis of encapsulated copper nanoparticles using a hydroalcoholic extract of moringa oleifera leaves and assessment of their antioxidant and antimicrobial activities. Molecules 25(3). https://doi.org/10.3390/molecules25030555
9. Díaz-Visurraga J, Daza C, Pozo C, Becerra A, Plessing C, García A (2012) Study on antibacterial alginate-stabilized copper nanoparticles by FT-IR and 2D-IR correlation spectroscopy. Int J Nanomed 7(2014):3597–3612. https://doi.org/10.2147/IJN.S32648
10. Naseer M, Ramadan R, Xing J, Samak NA (2021) Facile green synthesis of copper oxide nanoparticles for the eradication of multidrug resistant Klebsiella pneumonia and Helicobacter pylori biofilms. Int Biodeterior Biodegrad 159:105201. https://doi.org/10.1016/j.ibiod.2021.105201
11. Sarwar N et al (2021) Citric acid mediated green synthesis of copper nanoparticles using cinnamon bark extract and its multifaceted applications. J Clean Prod 292:125974. https://doi.org/10.1016/j.jclepro.2021.125974
12. Keabadile OP, Aremu AO, Elugoke SE, Fayemi OE (2020) Green and traditional synthesis of copper oxide nanoparticles—comparative study. Nanomater 10(12):1–19. https://doi.org/10.3390/nano10122502
13. Khan A, Rashid A, Younas R, Chong R (2016) A chemical reduction approach to the synthesis of copper nanoparticles. Int Nano Lett 6(1):21–26. https://doi.org/10.1007/s40089-015-0163-6
14. Lal SS, Nayak PL (2012) Green synthesis of gold nanoparticles using various extract of plants and spices. Int J Sci Innov Discov 2(3):325–350
15. Poopathi S, De Britto LJ, Praba VL, Mani C, Praveen M (2015) Synthesis of silver nanoparticles from Azadirachta indica—a most effective method for mosquito control. Environ Sci Pollut Res 22(4):2956–2963. https://doi.org/10.1007/s11356-014-3560-x
16. Shankar SS, Rai A, Ahmad A, Sastry M (2004) Rapid synthesis of Au, Ag, and bimetallic Au core Ag shell nanoparticles using neem (Azadirachta indica) leaf broth. J Colloid Interface Sci 275(2):496–502. https://doi.org/10.1016/j.jcis.2004.03.003

Development of a Method to Produce a Potential Transparent Conductive Material

B. R. Flores-Hernández and J. Santos-Cruz

Abstract A method for the preparation and characterization of graphene using 6 commercial detergents and sodium cholate by using two different equipment: sonication bath and a mix reactor is presented. For the experiment, a sonic bath with variable frequency and a beaker of 100 mL were used for testing. There were used different reactive varying its concentration, duration of the experiment, initial concentration of graphite, spin time, the centrifuge speed and frequency for sonication. Samples were taken and subsequently washed on a Buchner funnel with deionized water and methanol, and then be dried to store as powder. Finally, they were analyzed by UV–Vis to determine the concentration in the samples, Raman spectroscopy looking for crystallization patterns of graphene, TEM and HRTEM in order to determine the number of layers of graphite stacking. The detergents with the best perspective for applications were number 5 and 2.

Keywords Conductive · Shear exfoliation · Surfactants · Transparent · Graphene

1 Introduction

The discovery of graphene in 2004 by Novoselov and Geim has revolutionized the vision of nano materials [1]. Graphene has shown outstanding heat transfer, energy and hardness properties [2, 3]. Nowadays there are investigating methods to help mass production without damaging the environment. Methods that involve peeling in aqueous medium are developed technologies that have potential for industrialization. Mainly there are two methods that involve physical phenomena: sonication, which is in sonic bath, and the shear exfoliation, which is done in a container with propeller to generate turbulence [4].

Extensive efforts in this approach have been made to improve the yield and the quality of graphene. Organic solvent-based exfoliation, polymer-based exfoliation and surfactant-based exfoliation are three different approaches in liquid-phase

B. R. Flores-Hernández (✉) · J. Santos-Cruz
Facultad de Química, Materiales-Energía, Universidad Autónoma de Querétaro, C.P. 76010 Santiago de Querétaro, QRO, México

K. L. Flores Rodríguez et al. (eds.), *Recent Trends in Sustainable Engineering*, Lecture Notes in Networks and Systems 297,
https://doi.org/10.1007/978-3-030-82064-0_14

exfoliation method. Despite some defects brought by surfactant, the last one was proved to be an ideal way of preparing graphene dispersion with high graphene concentration, and more importantly, with excellent stability [5]. Ionic surfactants were first introduced to assist the exfoliation process. For instance, Vadukumpully et al. used a cation surfactant cetyltrimethylammonium bromide [6], Hernandez et al. used sodium dodecyl benzene sulfonate, sodium cholate and other ten kinds of surfactants to exfoliate graphite flakes [7]. Nuvoli et al. designed a series of works to get extremely high graphene concentration. Graphene concentration as high as 5.33 mg mL^{-1} is obtained in a commercial ionic liquid 1-hexyl-3-methyl-imidazolium hexafluorophosphate, 2.21 mg mL^{-1} in Nmethylpyrrolidone solution, 9.45 mg mL^{-1} in polymerizing media, and 8.00 mg mL^{-1} in organosilanes [8]. Du et al. introduced some organic salt to assist exfoliation and enhanced the exfoliation efficiency [9]. For non-ionic surfactant, Guardia et al. first explored the differences between ion-ic and non-ionic surfactants in assisting exfoliation and verified the ascendancy of non-ionic surfactant, and then extended the method to synthesize inorganic graphene analogues [10]. In addition, Niu et al. obtained graphene dispersion with enhanced graphene concentration with the assistant of inorganic salts. Wang et al. introduced ethanol into the surfactant/water solution to reduce exfoliation energy in surfactant/water medium and enhanced graphene concentration up to 3 times [11]. Samoilov et al. adopted an effective fluorinated surfactant for graphene production, which is environmentally friendly [12].

All the previous works show the advantage and the potential of surfactant assisted liquid-phase exfoliation method. The continuous research on in this field is thus necessary and meaningful [13]. To improve the approach of surfactants in exfoliation methods there are two significant problems to be considered. Firstly, what are the main factors that can influence the degree of exfoliation. Secondly, which parameters can represent the effectiveness of a method. For the first question, based on the predecessors works, some particular factors, for instance surfactant type, sonication time, centrifugation speed and initial graphite concentration were discussed as a function of graphene concentration. For surfactant type, Smith et al. proposed that, ionic and non-ionic surfactants have different mechanisms for stabilizing graphene dispersions. For sonication time, half hour sonication may have a decent marginal benefit over longer or shorter sonication time [14]. For centrifugation speed, the increase of centrifugation will negatively affect graphene concentration and graphene sheet's quantity. However, to the best of our knowledge, the influence of surfactant concentration on graphene concentration have not been deeply explored. For the second question, Coleman et al. first used the dispersion absorption as a main index for exfoliation according to the Lambert–Beer Law, and used transmission electron microscope (TEM) and other characterization tools to examine the quality of the dispersion. Whether graphene dispersion with high graphene concentration has the same quality as the one with relatively low concentration is not solidly confirmed [15].

In this paper, six different of commercial detergents with different proportions of surfactants are used to give possible answers to these questions based on a file from PROFECO of 2014 [16]. The main reason for the use of those detergents is because are environmental friendly and reduce the impact of pollution. The optimum concentrations of reactive for exfoliation were found. In order to get a full understanding of the factors that influence graphene concentration, many controlled experiments are carried out. Characterization methods are performed to examine the quality (sheet size, number of the layers and structural defects) of the product. The results provide valuable data and references for graphene exfoliation in water/surfactant dispersion. The objective of the research was to synthesize graphene using biodegradable detergents by two methodologies: Shear exfoliation and sonochemistry. They were tested and with them determining the quality of the graphene obtained by Raman spectroscopy, UV–Vis, TEM and HRTEM.

2 Methodology

A mixture of graphite flakes from Sigma-Aldrich, deionized water and 6 commercial products with different surfactant concentration were used, additionally Sodium cholate was used as a contrast patron. All reactive were mixed in a polymer reactor of 2 L with two blades of 7 cm, enhanced the turbulence in the container made in lab. There was performed a range of mixing experiments varying a range of parameters: Initial concentration of graphite, blade speed, type of reactor and mixing time. The mixer was operated at 25,000 rpm for 45 min. The engine of the reactor is not designed for continuous operation at high speeds for long times due to excess heating. After the mixing, aliquots of the resultant dispersions were collected (5 mL) and centrifuged at 1500 rpm for 45 min.

The methodology used for the method of the samples in sonic bath together with exfoliation shear were as follows:

The amount of surfactant for each of the runs were weighed and mixed first with deionized water for a period of 5 min so that the concentration in the sample was homogeneous. Subsequently graphite flakes were added. The procedure was the same for the reactor and for sonic bath, with the variation that the reactor volume used was 500 mL. For the procedure for the sonic bath, it allowed for a period of 82 min. Batch was evaluated with 4 samples, which were equidistant from the center. In the case of the reactor, the methodology involves testing a minute in the reactor on the other in an ice bucket, because the heat of the blades is transferred to the mixture of graphite with water and may actually cause evaporation mixture and this will cause change in concentration. Temperature plays a role in the surface tension, therefore, the smaller the variation in temperature, better control of the properties of the mixture, then the surface tension to be constant is controlled [7].

To make an energetic comparison between the two methods, the direct relationship between the equipment of shear exfoliation and sonic bath was made. The amount of energy used in shear exfoliation equipment were 1400 W. With that amount of energy for the process in the reactor, a period of 30 min was used. Then 2.5 MJ were the amount of energy consumed, to make an equivalent estimation in the sonic bath reactor, the work time was estimated using the power of 550 W, then to meet the same amount of joules supplied, the equipment must to have continuous operation for a period of 82 min to provide the same amount of energy.

One of the advantages of sonic bath process is that can work continually, it means that, once the process started it did not require supervision. The process inside both reactors suffered heating in the mixture at 48 °C, measurements were made on the samples and that was the maximum temperature reported for the process. Instead, the process reactor has to be intermittent as it undergoes heating in the motor can reach the boiling point of water, causing problems in the mixture, so was one minute turned on for one minute in ice bath. In both processes it has been observed that the average temperature throughout the process remained around 35 °C.

Exfoliation in the sonic bath was carried out as specified in the methodology. In general, to identify each of the samples was assigned a number to the level of each factor, the first number indicates the surfactant used, numbered 1–7. The second number indicates the concentration that was used, based on the Table 2, and goes from number 1–3, where 1 is the lowest concentration of surfactant and 3 the greatest. The last number indicates the level in the sample, for sonic bath was operated level 1 indicates 75% power the equipment and two 100% power. To the reactor level 1 is referred to the speed computer 2 and level 2 to speed 3.

3 Results

6 different commercial detergents (CD) with distinct composition were evaluated and Sodium Cholate. First, a relationship between concentration of commercial detergent and surface tension had to be found. It was reported that a surface tension of 46.7 mN m^{-1} [17] was necessary to exfoliate graphene from graphite in solvents. N-methyl-Pirrolidine (NMP) is the solvent with the best properties to exfoliate graphene because of its inherent surface tension at room temperature. The concentration of the CD has to achieve the values of NMP to provide a well exfoliated graphene.

In Table 1 is showed the different concentrations that were tried to achieve the surface tension that was reported to be the best for graphite exfoliation. The method

Table 1 Surface tension of each type of detergent related to its concentration

Detergent	Level of concentration of detergent	Level of power	Concentration of graphene [mg/mL]
1	3	2	0.053
2	3	2	0.086
3	3	1	0.081
4	3	2	0.052
5	3	1	0.086
6	3	1	0.069
7	3	2	0.070

Table 2 High concentration of graphene for sonic bath of each type of detergent

Detergent	Concentration [mg/mL]	Surface tension [mN/m]
1	0.05	48.45
1	0.1	41.99
1	0.15	47.4
2	0.05	47.75
2	0.1	48.01
2	0.15	47.49
3	0.05	47.06
3	0.1	46.7
3	0.15	48.45
4	0.05	53.25
4	0.1	43.21
4	0.15	48.45
5	0.05	48.36
5	0.1	50.45
5	0.15	54.99
6	0.05	48.45
6	0.1	45.65
6	0.15	43.65
7	0.05	44.95
7	0.1	46.44
7	0.15	46.7

used for this purpose was the Du Nouy ring. It consists in the use of a wheel to measure the force to bring out a ring from the surface of the liquid contained.

With the data obtained, the concentration was estimated and then the process to evaluated were sonic bath and shear exfoliation. In Table 2 the values of the highest concentration obtained for each CD are presented (0.15 mg mL^{-1}), based in the relation between concentration and surface tension, the highest value of concentration was used because the value of surface tension was close to that reported for sodium cholate. There are also presented the power used for the production of graphene and the final concentration of graphene obtained.

The absorbance was analyzed with the UV–Vis equipment, in order to determine the concentration, the measurements were performed from 190 to 800 nm. In Table 3 are the data obtained by sonic bath of the highest concentrations for each of the detergents, the detergents 2, 3 and 5 are those with higher concentrations of graphene, while the others had a deficient performance, It is important to emphasize that is sought is a higher concentration of graphene based on the wavelength of 660 nm. Of the reagents with the highest concentration, those with the best results were 2 and 5 with a value of 0.086 mg mL^{-1} surpassing that found for sodium cholate, represented by reactive 3. In the case of the results obtained by shear exfoliation

Table 3 High concentration of graphene for shear exfoliation of each type of detergent

Detergent	Level of concentration of detergent	Level of power	Concentration of graphene [mg/mL]
1	2	1	0.011
2	3	1	0.035
3	2	2	0.040
4	2	1	0.012
5	3	2	0.026
6	3	1	0.019
7	2	1	0.012

(Table 3), a similar behavior was presented regarding the 3 products with better yields, however, sodium cholate had a higher performance in terms of obtaining graphene compared to detergents proposed 2 and 5. From the two methods used, the one were higher concentration values were obtained was sonic bath, however it has been reported that the technique is not as scalable as the shear exfoliation, hence the interest to study and compare both.

In the case of Raman spectroscopy for sonic bath there is a match with the results. Two more samples of reagents 2 and 5 were also analyzed since the other levels of the treatments also had graphene concentrations similar to the highest. Figure 1 shows the spectra of each of the samples analyzed, D (1350 cm^{-1}), G (1583 cm^{-1}) and

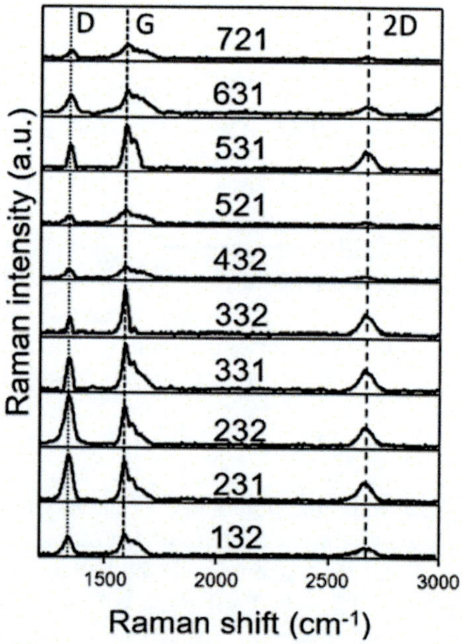

Fig. 1 Raman spectroscopy of graphene obtained by different treatments of sonic bath exfoliation

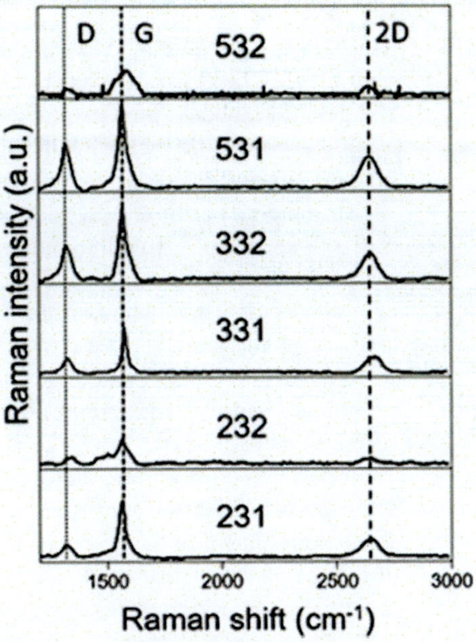

Fig. 2 Raman spectroscopy of graphene obtained by different treatments of shear exfoliation

2D (2680 cm^{-1}) bands are signposted with dash lines. For sample 231 and 232, no significant difference was seen in terms of intensities of D and G band, the presence of other bands or a spectral shift, but in the case of samples 331 and 332 this change was similar. The sample 332 does not present the band D′ related to the lack of impurities, however a greater value is presented as regards the relation of the bands D with G, resulting in larger sheet sizes.

Finally, for the detergent 5, the sample 531 presents a spectrum similar to the detergent 2, which depicts of the presence of impurities but which it is still possible to use.

In the case of shear exfoliation in the Raman spectra (Fig. 2), in general, the lack of G′ band is observed, indicating a lack of impurities. A determining factor to take into account is the ratio of the D and G bands, which presents the largest values of the analyzed samples are those belonging to the detergent 5, indicating a better graphene quality obtained. Supported by the intensity of the spectra. This D/G ratio is comparable to that observed by the sonic bath method but with better results since the band G′ belonging to the impurities is not shown.

Although Raman spectroscopy is a useful for determining hybridization of sp2 and sp3 in carbon atoms. The presence of D (1350 cm^{-1}), G (1583 cm^{-1}), the D ‘(1620 cm^{-1}) and 2D (2680 cm^{-1}) are "footprint" of graphene. Each band gives different information. D band is caused by a disorder in the structure of graphene. The presence of disturbances in the system sp2 hybridized carbon is showed in the band with high intensity. Therefore, Raman spectroscopy is a technique used to determine the hybridization of a material. In a perfect structure of graphene D band is almost imperceptible. The G band is due to a mode E2g in gamma dots. The G band arises by stretching the carbon graphite, which makes common systems sp2 hybridized. The G band is sensitive to stress [18]. If there is any impurity or surface charges in graphene, the G band can be divided into two peaks, the G and D′. The main reason for this behavior in materials is due to phonon interaction, causing a division in bands [19].

All materials with sp2 hybridized carbon, exhibit strong intensity in the range of 2500–2800 cm^{-1}. In combination with band G becomes a foot print of graphitic materials sp2 hybridized. G′ or 2D′ band is given by the process of phonon interaction of second order, this behavior has a strong frequency dependence of the laser used [20].

In addition, the 2D band can be used to estimate the number of graphene layers present in a sample. The shape of the 2D band single layer graphene is different from a multilayer sample.

In order to corroborate, the observed by Raman spectroscopy, High Resolution Transmission electron microscopy (HRTEM) was used. In Fig. 3, the sample 231 **shows four images of the graphene sheet at different magnification (left side), the area** of the sheets is superior of the 1 μm^2, and when analyzed the diffraction pattern for the same sample (right side), each line represents, as Hernandez et al. suggested, that the brightness of the inner dots most be bigger than the outsiders. This are represented with the graphics below the diffraction pattern. If the inner picks are greater, then the graphene sheet most have one or two layers, like in this case.

On the other hand, the analyses for the sample with the higher quality for shear exfoliation according to the Raman spectra was 531. In the left side there are 4 images **with different magnification. In image (a) and (b) there are graphene sheets with are** equal to sonic bath, but in (a) there is a roll of a sheet what could it be inferred as a carbon nanotube. Images (c) and (d) depict section of sheet with planes that were analyzed with a diffraction pattern. On the right side of Fig. 4, there are 3 graphics

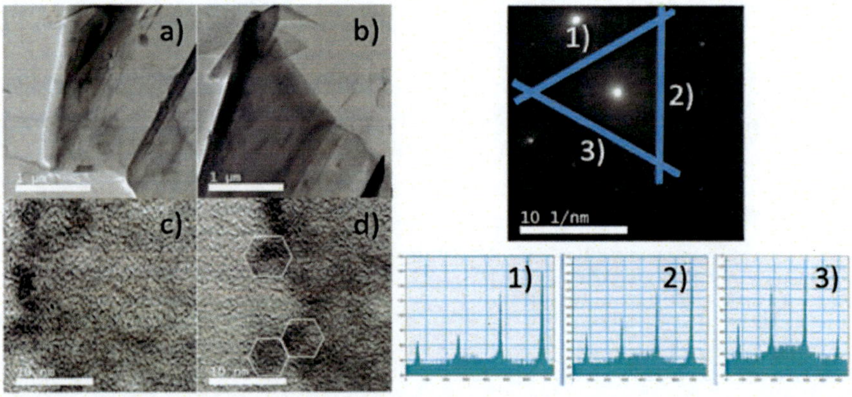

Fig. 3 231 sample of shear exfoliation. (Left side), TEM of the sample. (Right side), diffraction pattern and intensity graphs

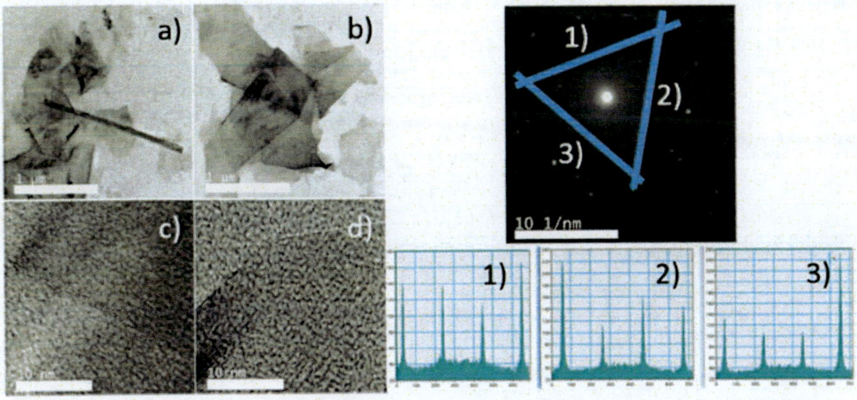

Fig. 4 531 sample of shear exfoliation. (Left side), TEM of the sample. (Right side), diffraction pattern and intensity graphs

below the diffraction pattern that represent each of the lines crossing. In this case, the inner picks are shorter than the outsiders meaning that the graphene sheets are stacking with more of 5 layers.

In the sample 531 there is a stacking effect. The stacking effect is the superposition of sheets of graphene because of it is in suspension. There are two types of possible order of the stacking: the ABAB and the ABCAB as it depicts the Fig. 5.

The order of the layers modifies material properties as thermal or electric conductivity. The ABAB stack consists in the superposition of sheets of graphene in parallel. While, ABCAB arrangement consists in stacking sheets covering the vacancies of space, forming the graphite cluster [5].

The second form is weaker than the first and the properties are more like an insulator. According to Hernandez, after 10 layers the stack of carbon sheets is in an ABCABC form. This is the reason that after 10 layers of carbon the properties of graphene change into graphite.

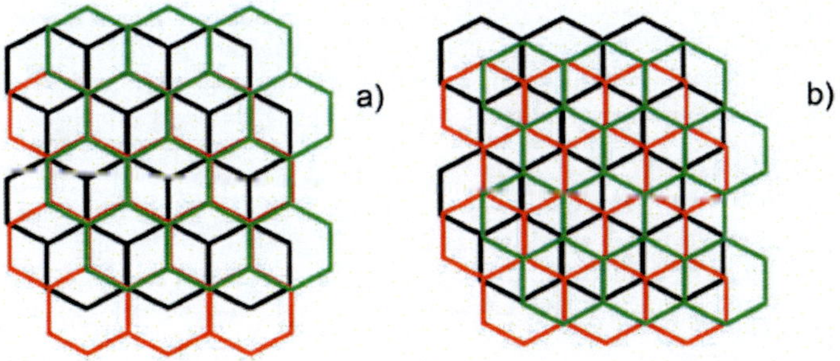

Fig. 5 Layer arrangement of 3 stack graphene sheets. **a** Represents ABAB stack and **b** represents the ABCAB stack

4 Conclusion

It has been demonstrated that graphite can be exfoliated to give few layers graphene using a simple methodology of rotating-blade mixer reactor. Moreover, sophisticated surfactants as sodium cholate are not necessary to stabilize the exfoliated graphene, commercial detergent works extremely well. It has been found that, the concentration of exfoliated graphene increases linearly with time, resulting in a time-independent production rate. In addition, concentrations of 1 mg mL^{-1} were achieved, it was higher with rotor–stator mixers than what can be achieved with sonication. It is suggested that exfoliation is enabled by the locally high shear rates associated with high Reynolds number turbulence. With the results obtained, there is the prospect of manufacturing thin films of this material that allow it to be used in optoelectronic devices.

Acknowledgements National Council of Science and Technology (CONACYT) for their support in the master thesis of Bruno Renato Flores Hernández.

References

1. Novoselov KS, Geim AK, Morozov SV, Jiang D, Zhang Y, Dubonos SV, Grigorieva IV, Firsov AA (2004) Electric field effect in atomically thin carbon films. Sci 306:666–669. https://doi.org/10.1126/science.1102896
2. Zhu Y, Murali S, Cai W, Li X, Won S, Potts J, Ruof R (2010) Graphene and graphene oxide: synthesis, properties, and applications. Adv Mater 22:3906–3924. https://doi.org/10.1002/adma.201001068
3. Walker M, Weatherup R, Bell N, Hofmann S, Keyser U (2015) Free-standing graphene membranes on glass nanopores for ionic current measurements. Appl Phys Lett 106:023119–023121. https://doi.org/10.1063/1.4906236
4. Coleman JN (2009) Liquid-phase exfoliation of nanotubes and graphene. Adv Funct Mater 19:3680–3695. https://doi.org/10.1002/adfm.200901640
5. Hernandez Y, Nicolosi V, Lotya M, Blighe F, Sun Z, De S, Mcgovern I, Holland B, Byrne M, Gun'ko Y, Boland J, Niraj P, Duesberg G, Krishnamurthy S, Goodhue R, Hutchison J, Scardaci V, Ferrari A, Coleman J (2008) High-yield production of graphene by liquid-phase exfoliation of graphite. Nat Nanotechnol 3:563–568. https://doi.org/10.1038/nnano.2008.215
6. Vadukumpully S, Paul J, Valiyaveettil S (2009) Cationic surfactant mediated exfoliation of graphite into graphene flakes. Carbon 47:3288–3294. https://doi.org/10.1016/j.carbon.2009.07.049
7. Varrla E, Paton K, Backes C, Harvey A, Smith R, McCauleyac J, Coleman J (2014) Turbulence-assisted shear exfoliation of graphene using household detergent and a kitchen blender. Nanoscale 6:11810–11819. https://doi.org/10.1039/C4NR03560G
8. Nuvoli D, Valentini L, Alzari V, Scognamillo S, Bon S, Piccinini M, Illescasd J, Mariani A (2011) High concentration few-layer graphene sheets obtained by liquid phase exfoliation of graphite in ionic liquid. J Mater Chem 21:3428–3431. https://doi.org/10.1039/C0JM02461A
9. Du W, Lu J, Sun P, Zhu Y, Jiang X (2013) Organic salt-assisted liquid-phase exfoliation of graphite to produce high-quality graphene. Chem Phys Lett 568:198–201. https://doi.org/10.1016/j.cplett.2013.03.060

10. **Guardia L, Paredes J, Rozada R, Villar-Rodil S, Martınez-Alonso A, Tascon J (2014) Production** of aqueous dispersions of inorganic graphene analogues by exfoliation and stabilization with non-ionic surfactants. RSC Adv 4:14115–14127. https://doi.org/10.1039/C4RA00212A
11. Niu L et al (2013) Salt-assisted direct exfoliation of graphite into high-quality, large-size, few-layer graphene sheets. Nanoscale 5:7202–7208. https://doi.org/10.1039/C3NR02173D
12. **Samoilov VM et al (2015) Formation of graphene aqueous suspensions using fluorinated** surfactant-assisted ultrasonication of pristine graphite. Carbon 84:38–46. https://doi.org/10.1016/j.carbon.2014.11.051
13. Amiri A, Naraghi M, Ahmadi G, Soleymaniha M, Shanbedi M (2018) A review on liquid-phase exfoliation for scalable production of pure graphene, wrinkled, crumpled and functionalized graphene and challenges. FlatChem 8:40–71. https://doi.org/10.1016/j.flatc.2018.03.004
14. Smith RJ, Lotya M, Coleman JN (2010) The importance of repulsive potential barriers for the dispersion of graphene using surfactants. New J Phys 12:125008–125011. https://doi.org/10.1088/1367-2630/12/12/125008
15. Paton KR et al (2014) Scalable production of large quantities of defect-free few-layer graphene by shear exfoliation in liquids. Nat Mater 13:624–630. https://doi.org/10.1038/nmat3944
16. Revista del Consumidor Profeco (2014) 452:88–98
17. Lotya M, King PJ, Khan U, De S, Coleman JN (2010) High-concentration, surfactant-stabilized graphene dispersions. ACS Nano 4:3155–3162. https://doi.org/10.1021/nn1005304
18. Pimenta M et al (2007) Studying disorder in graphite-based systems by Raman spectroscopy. Phys Chem 9:1276–1290. https://doi.org/10.1039/B613962K
19. Jorio A, Saito R, Dresselhaus G, Dresselhaus MS (2011) Raman spectroscopy in graphene related systems. Wiley
20. Ferrari AC, Meyer JC, Casiraghi C (2014) Optoelectromechanical multimodal biosensor with graphene active region. Nano Lett 14:5641–5649. https://doi.org/10.1021/nl502279c

Mechatronic Design Methodology for Fast-Prototyping of a Pressure Controlled Mechanical Ventilator

Fernando Martell, Jorge Mario Uribe, Juan Sarabia, Armando Ruiz, Ángel Eugenio Martínez, and Eduardo Licurgo

Abstract The Mechanical Ventilator to support human breathing is a medical device that became essential for the attention of patients with respiration complication due to the Covid19 pandemic disease. The rapid prototyping and development of such devices was encouraged at academic and research institutions around the world. This paper presents a mechatronic design methodology that was redefined and applied for the design and prototype of a Pressure Controlled Mechanical Ventilator. This work is presented as a case of study to show that good design practices such as model based design and "hardware in the loop" simulations, carried out in a methodological way, allows a faster and reliable development of mechatronic products.

Keywords Mechanical ventilator · Mechatronic design methodology

1 Introduction

The unfortunate expansion of the Covid19 pandemic disease raised concerns about the availability of mechanical ventilators at the hospitals attending the infected population with severe respiration complications. Educational institutions around the world shared designs and efforts to develop prototypes of mechanical ventilators. High expectations of the rapid prototyping of the medical device imposed to the research and development institutions the need for speeding up their design processes to provide mechanical ventilator designs for possible mass production. This was an opportunity to test the engineering design capabilities of these institutions.

Mechatronics is an interdisciplinary field, in which the following disciplines act together: (1) mechanical systems; (2) electronic systems; (3) information technology. The integration of mechanism with electronics devices and information processing systems is performed within the hardware components (mechanism, sensor, actuators) and in the software functions (control and connectivity) [1]. The design process

F. Martell (✉) · J. M. Uribe · J. Sarabia · A. Ruiz · Á. E. Martínez · E. Licurgo
Centro de Investigaciones en Óptica, A.C., Aguascalientes, México
e-mail: fmartell@cio.mx

K. L. Flores Rodríguez et al. (eds.), *Recent Trends in Sustainable Engineering*, Lecture Notes in Networks and Systems 297,
https://doi.org/10.1007/978-3-030-82064-0_15

has to be concurrent, since "the design of the mechanical system influences the electronic system and the design of electronic system also has influence on the mechanical system" [2].

The application of a good mechatronic methodology should result in a better product optimized in cost, performance and fabrication time. By reviewing the mechatronic design process and methodologies in literature [1–5], it is found the recurrent recommendation to use good design practices such as "model based design" and "real time simulation" for better product design and integration. In the case of a mechanical ventilator, prior to develop a prototype it is convenient to reproduce the behavior of artificial ventilation process for a better understanding of the critical functions and the system dynamics. There are research works about the mathematical modelling and computational simulation of a mechanical ventilation process, among them [6–9], they offer complementary approaches to gain knowledge about the mechanical ventilation systems.

This paper presents a mechatronic design methodology based on mathematical modelling and "hardware in the loop simulation" of a Pressure Controlled Mechanical Ventilator (PCMV). The paper is organized as follows: Sect. 2 presents the operating principles of mechanical ventilators and fundamental concepts for their modelling and simulation. In Sect. 3, a brief review of some conventional mechatronic design methodologies is discussed and the herein proposed mechatronics design methodology is presented. Section 4 describes the design and prototype of the mechanical ventilator as a case of study and presents results of some performed functional and operational tests. Section 5 is of conclusions.

2 Mechatronic Design Methodologies

2.1 Mechatronic Design Process

The design methodologies for the development of mechatronic products are relevant because they provide means to better organize and execute the diverse design and development tasks such as: specification, design, fabrication, prototyping, integration and validation and verification tests. The mechatronic design process according to [3] consists of three phases: modeling and simulation, prototyping, and deployment as depicted in Fig. 1.

Mechatronic products are comprehensive mechanical systems with fully integrated electronics and information technology (IT). Such systems require another approach for efficient development as pure mechanical, electronic/electric and IT products [5]. VDI 2206, see Fig. 2, is a widely accepted industrial guidelines that defines crucial steps and measures to finalize efficient and cost-efficient mechatronic products. The objective of this guideline is to provide methodological support for a cross-domain development especially in the early phase of development, concentrating on system design [5]. As a whole, the guideline consists of three essential

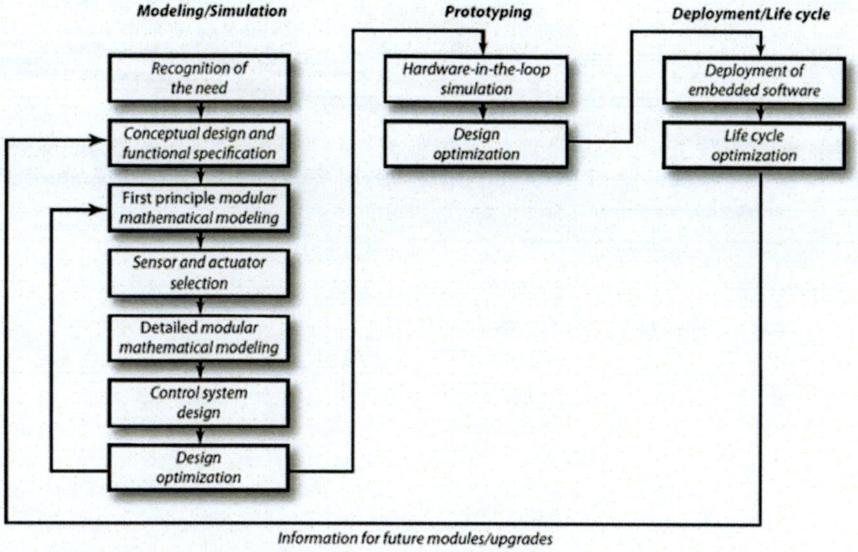

Fig. 1 Mechatronic design process from Shetty and Kolk [3]

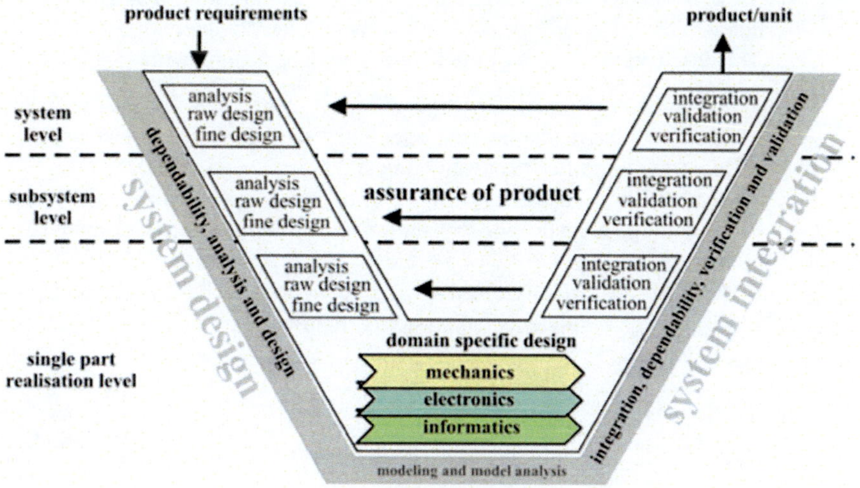

Fig. 2 V-model for mechatronic product development [5]

elements: a general problem-solving cycle as a micro-cycle, the V-model as a macro-cycle, and predefined process modules for recurrent working steps. In the description of the micro-cycle, the guideline VDI 2206 refers to the problem-solving cycle used in systems engineering.

The mechatronics design process of [3] and the VDI 2206 methodology consider the employment of model based design and hardware in the loop simulations as a part of the design process. VDI 2206 also specifies the need for validation and verification during the system integration, at each system, subsystem and component-level. The objective of verification and validation is to gain confidence in the correctness of a system with respect to its specification (verification) and when it is placed into its target environment (validation).

2.2 Proposed Methodology

The proposed methodology depicted in Fig. 3 is based on the V-Model and takes important concepts and considerations from the VDI2206 standard, this methodology reinterprets the design process and rearranges the stages in such manner that imposes a logical continuity and connection among them. It recognizes the need to define both concurrent and sequential activities that implies precedence and subsequence. The three levels represent a hierarchy, at the higher level are the system design and integration inputs and outputs, the second level is software centered and the lower level is related to the hardware components. The activities at the left are related with the design and at the right are related with the integration, such as in the VDI2206 standard. This methodology is intended for fast development of a system prototype.

The description of each of the herein proposed design methodology stages are:

(1) System Requirement and Specification: This entry stage is user and application centered, defines the user interface and embodiment considerations. Identification of critical functions and process variables;
(2) Model Based System Design: Functional Specification, CAD based 3D modelling, Signals and System response modelling and analysis, Bill of materials;
(3) Acquisitions and Fabrication: Components purchase and procurement, By discipline components (subsystems) Fabrication;

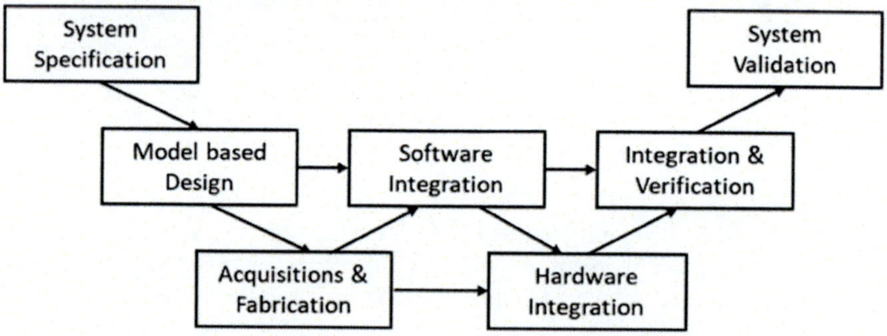

Fig. 3 Proposed mechatronic design process

(4) Software Integration: Software or preferably Hardware in the loop Simulation (to verify modelling and control), Connectivity and User Interface functional test;
(5) Hardware Integration: Component Level integration of the hardware devices such as the mechanism, sensors, actuators;
(6) System Integration and Verification: Integration of Hardware and Software Components; Verification of Functional Specifications; and,
(7) System Validation and Deployment. Target environment test. Functional and operational validation.

The proposed methodology was redefined to identify the best path for inputs and outputs of the different stages of the design process, in this sense, it is and interpretation of the main concepts of the VDI-2206 but simplified and applied for the fast design of a system prototype. An advantage of this design methodology is that each stage has clearly defined outputs that are inputs to the next stage as it is shown in Table 1.

Table 1 Proposed methodology stage and outputs

Stage	Output
(1) System requirements and specification	Functional specification/description
(2) System model based design	2D and 3D CAD mechanical drawings ECAD electronic drawings Signals and systems model Bill of materials
(3) Acquisitions and fabrication	Mechanical fabricated components Control panel and electronic PCBs
(4) Software integration	Control system verification test
(5) Hardware integration	Hardware verification test
(6) System integration and verification	System functional test (verification)
(7) System validation and deployment	System operational test (validation)

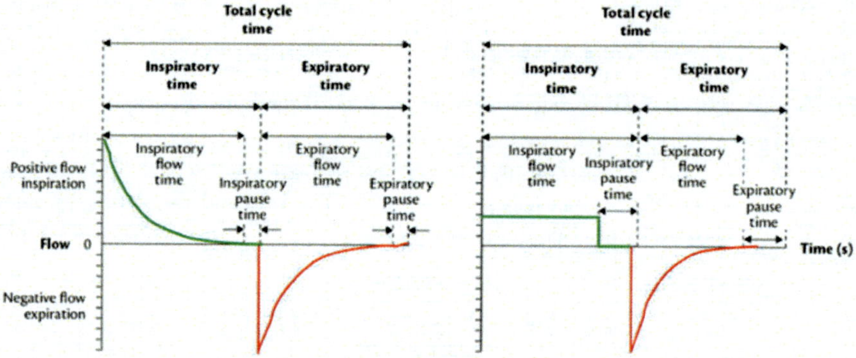

Fig. 4 (Left) pressure control ventilation. (Right) volume control ventilation [10]

3 Mechanical Ventilation Systems

3.1 *Mechanical Ventilator Operation and Functions*

The mechanical ventilator is a mechatronic machine designed to move air in and out of the lungs for patients requiring support for breathing. The functional specification of a ventilator implies the understanding of how must be provided the work of breathing to a patient that is a function of pressure and volume; either the patient's inspiratory muscles or the mechanical ventilator generates an increase in the pressure difference across the lungs [10].

There are two main types of mechanical ventilation, these are: Pressure Control Ventilation (PCV) and Volume Control Ventilation (VCM), see Fig. 4. A breath is defined as one cycle of inspiratory flow followed by a matching expiratory flow [10]. An assisted inspiration can be recognized as one for which airway pressure rises above a baseline (positive end expiratory pressure) during inspiratory flow. On the other hand, if airway pressure drops below baseline pressure during inspiration, the patient is doing some work on the ventilator and the breath is 'loaded', rather than assisted [11].

3.2 *Modelling and Simulation of Mechanical Ventilators*

The mathematical model represents the respiratory activities and an important controlled parameter during mechanical ventilation, what it is called Positive End Expiration Pressure (PEEP). The Pressure signal over the complete respiratory cycle can be modeled in time domain as in Eq. (1).

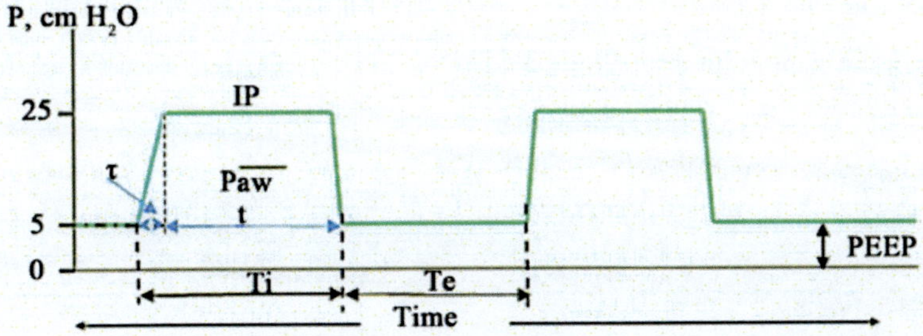

Fig. 5 Typical waveform of pressure signal for PCMV [8]

$$P(t) = \begin{cases} P_{aw} \cdot \frac{t}{\tau} + PEEP, & 0 \leq t \leq \tau \\ P_{aw} + PEEP, & \tau \leq t \leq T_{in} \\ PEEP, & T_{in} \leq t \leq T_{ex} \end{cases} \quad (1)$$

where

$P(t)$	is the pressure signal of the PCV;
$PEEP$	is the positive end-expiratory pressure;
P_{aw}	is the pressure in respiratory airway;
τ	is a rise time.
T_{in}	is the inspiration time, and;
T_{ex}	is the expiration time.

Asymmetrical pulse periodic functions defined in previous Eq. (1) are used to control the inspiration and expiration durations, see Fig. 5, [8], and therefore to drive the mechanical ventilator.

Computational routines with parameter based estimation methods can be utilized to estimate PEEP, resistance and compliance of the artificial respiratory system [8]. A multi-compartment model for the human lungs based on an electric circuit was first suggested by Michael [9], see Fig. 6, this model can estimate the process variables of an artificial respiratory system.

Fig. 6 Multi-compartment electrical model [9]

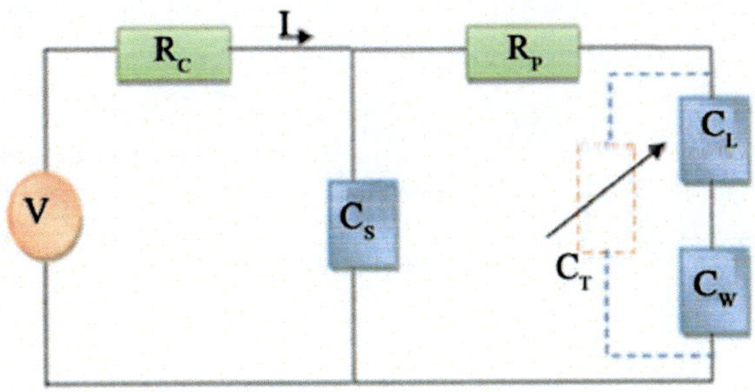

The current flow, I, represents airflow, and the voltage source, V, corresponds to the applied pressure produced by the ventilator, the electric components of the circuit are:

- $R_C = 1$ cm H_2O/L/s, is the airflow resistance of the central airways.
- $R_p = 0.5$ cm H_2O /L/s, is the resistance of the peripheral airways.
- $C_L = 200$ ml/cm H_2O, is the capacity of the alveoli.
- $C_W = 200$ ml/cm H_2O, is the chest wall capacity, which is in series with the alveoli.
- $C_S = 5$ ml/cm H_2O is a shunt capacitance known as "dead space" of air, which does not participate in the exchange of oxygen and carbon dioxide between air and blood.
- C_T is the total compliance of airways that depends on C_L and C_W.

This circuit model has been converted to a mathematical model using a transfer function [8], the input–output transfer function in frequency domain, Eq. (2) reproduces the dynamical response of the airflow (output) as a function of input pressure.

$$\frac{I(s)}{P(s)} = \frac{s^2 + \frac{s}{R_p \cdot C_T}}{R_c s^2 + \left(\frac{1}{C_s} + \frac{R_C}{R_p \cdot C_T}\right)s + \left(\frac{1}{R_p \cdot C_s}\right)\left(\frac{1}{C_L} + \frac{1}{C_W}\right)} \quad (2)$$

With all model parameters adjusted for a normal lung of a human being, the transfer function to model the pressure output from the PMV is given in Eq. (3).

$$\frac{I(s)}{P(s)} = \frac{s^2 + 420s}{s^2 + 620s + 4000} \quad (3)$$

The respiration cycle of Eq. (1) to generate a voltage input proportional to a pressure waveform, has been applied to the multi-compartment model for lungs of Eq. (3), [9], to reproduce the behavior of the critical variables of the PCMV. Other modelling approaches based on pneumatic systems and state space representation of the mathematical models can be found in [6, 7]. There are recent works showing the advantageous of employing a V-model for the design of mechanical ventilators which also considers modeling and simulation stages [12].

4 Ventilator Prototype Design and Development

4.1 Model Based System Design

One of the first steps in any mechatronic design process is to identify the critical variables which are connected with the system functional specification, these process and control variables are required for the proper selection of sensors and actuators, and for the specification and design of the control system. A research in literature about previous modelling efforts of the system under design is needed to choose an appropriate mathematical model that can be simulated computationally for a better understanding of the behavior of the system and their inputs, outputs or any other state or process variables.

The computational simulation provides means for testing and monitoring the behavior of signals and other parameters (volume, flow and dynamic compliance) during artificial ventilation [5]. In this work the dynamical response of the pressure, air flow and tidal volume in continuous time domain (Sect. 3.2), were simulated in Simulink, by a pulse train for the pressure signal and the flow was modelled by a second order input–output transfer function considering the pressure as input. Figure 7 shows the Simulink diagram and the behavior of the simulated process variables. The simulation of the mathematical model allows also to determine the operational ranges of both process variables measured from sensors and control variables to be applied to actuators.

The simulation of the mathematical model allows also to determine the operational ranges of both process variables measured from sensors and control variables to be applied to actuators. At this point a definition of the control system can be done.

The variables identified in the model based design are now defined to be inputs and outputs to the electronic control unit, in this case a microcontroller, as shown in Fig. 8. The sensor inputs are for pressure, two photo-sensors to detect the ventilation semi-cycles and to compute the respirations per minute (RPM). A potentiometer to set RPM. The outputs are a PWM signal to control the speed to the Direct Current (DC) motor and a signal to display data in a Liquid Crystal Display (LCD).

For the correct selection of sensors, actuators and electronic devices for signal conditioning, it is important to determine the maximum and minimum values for the variables. These values will allow choosing the best sensor or actuator and other electronic components for electronic design. Table 2 shows the sensors and electronics devices for signal conditioning and amplifications for each variable based on the minimum and maximum values.

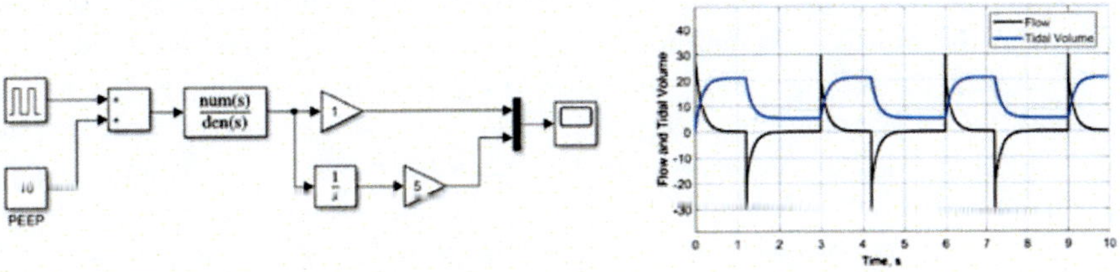

Fig. 7 Simulink block diagram (left) and the simulated process variables (right)

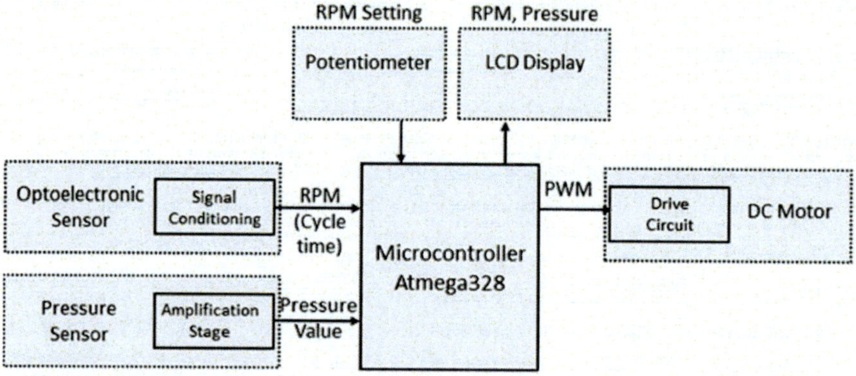

Fig. 8 Inputs and outputs of the control systems

Table 2 Maximum and minimum variables values

Variable	Minimum	Maximum	Sensor or electronic device
Signal for speed control	0 V	5 V	Potentiometer
RPM measure	1 RPM	30 RPM	CNY70
Pressure measure	10	50	MPX100DP
PWM output	1 kHz	10 kHz	Not apply

4.2 CAD Based System Design

The system design of a modern mechatronic systems is based on the use of a computer aided environment and should be done concurrently by discipline: Mechanical, Electronic, Controls and application software. Computer Aided Design (CAD) is a standard practice for the design of mechanical systems, each of the parts, components and subassemblies can be designed with proper CAD tools.

The PCMV under study was completely designed in SolidWorks, and it was even possible to make a 3D animation for better analysis of the dimensions and clearances among the mechanical parts. Some of the design intentions for the development of the PCMV were the following: Parts easy to obtain or manufacture, simple to assemble, to have the least number of adjustments for its operation, resistance and durability. Following the design intentions, the mechanical system was divided into three subsystems: the motor drive and gearing system, the ambubag compression system and the general structure. One of the mechanical features is the manually adjustable position of the system plungers for the compression of the ambubag, this allows to modify the pressing force, which directly modifies the volume of airflow.

Electronic Computer-Aided Design (ECAD), also referred to as Electronic Design Automation (EDA), is the principal tool for manufacturing integrated circuits. The

EDA software allows the developing of products more quickly, complex and efficient. The layout design process is the step that allows placement of components, define the rules for reliable manufacturing of the PCB and the essential step, definer routing (including sizes of traces for different currents). A PCB was designed to be a shield of the Arduino microcontroller; this is described in the section below.

One of the main outputs of the system design stage is the completion of a bill of materials that considers components and materials for the prototype fabrication that considers also the control systems and the required sensors actuators and electronic interface and signal conditioning components.

4.3 *Prototype Hardware and Software Integration*

The mechanical ventilator was fabricated as a robust device made of aluminum to ensure its mechanical stability, simple to machine and assemble; It has a paddle system which presses the ambubag in a controlled way, thus allowing the supply of air pressure and flow to the patient that can be configured for a range of RPMs. The periodic movement of the paddles is achieved by means of a DC motor which is coupled to a gear train and is controlled by photo sensors which allows to regulate the respiratory cycle. The control system is implemented in an Arduino microcontroller that is capable to optionally communicate with a LabVIEW interface by a serial link to monitor the RPM and the pressure. An Arduino shield was designed and fabricated, see Fig. 9, for signal conditioning an interface to the sensors and actuators.

The real time hardware in the loop simulation (HiL) is a powerful practice for software integration, useful for the verification of the control system. The mathematical model already validated in the design stage is converted to a discrete-time model using discretization methods such as Euler´s backward differences or bilinear transform (Tustin method). The discrete-time model is used for numerical simulation

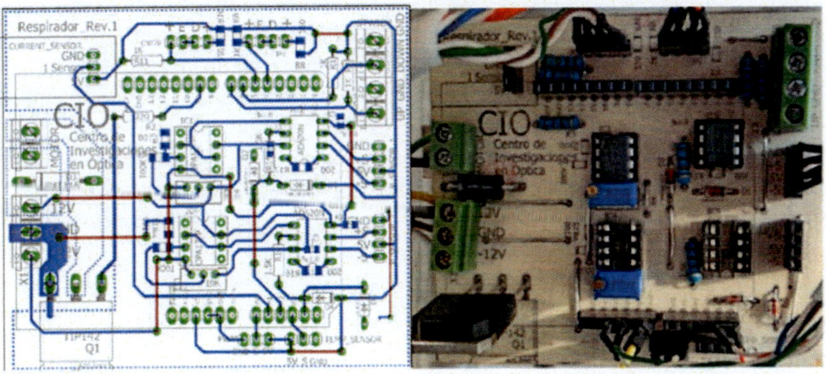

Fig. 9 PCB design and manufacture

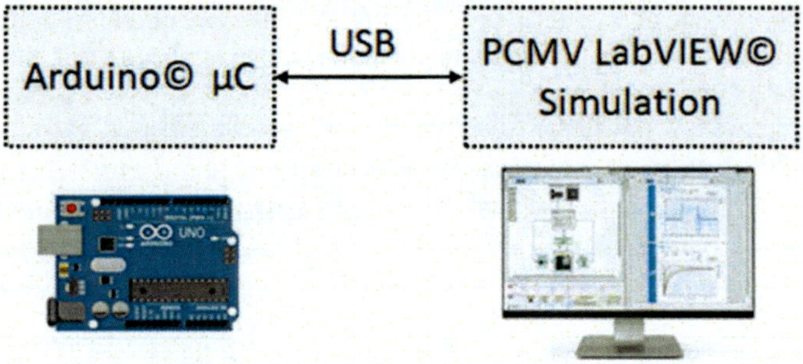

Fig. 10 Hardware in the loop simulation

of the process, this was implemented in LabVIEW to reproduce the process variables to the controller by responding to its output in closed loop scheme, see Fig. 10. In the proposed methodology, the HiL simulation is very convenient for the software integration and the verification. The performed HiL simulation in this project reproduced the pressure signal in discrete time described by [8] and the transfer function of the electrical circuit to model the lung proposed by [9].

4.4 System Integration, Validation and Verification

The overall system integration and the validation and verifications test are performed in this stage. The final test is with an artificial lung as is depicted in Fig. 11, the image at the left shows the ambubag at the inspiratory semi-cycle and the image of the right shows the ambubag finishing the expiratory semi-cycle. The system prototype of the PCMV was test during long periods of time to verify and validate the precision and repetition of the respiratory cycle.

Due to the large process for the accreditation of the PCMV for their real medical use, its application was initially considered to be for educational use, suitable for training ventilator functions and operations at medical schools. The system was tested at a medical laboratory from a university and the PCMV performed the respiratory support correctly, this was considered the final validation test of the medical device. Figure 12, shows the pressure waveform for the minimum respiratory cycle of 18.6 rpm and for the maximum at 27 rpm.

5 Conclusions

The design and fabrication of the prototype of a Pressure Controlled Mechanical Ventilator presented in this paper employed design practices like: 3D CAD based

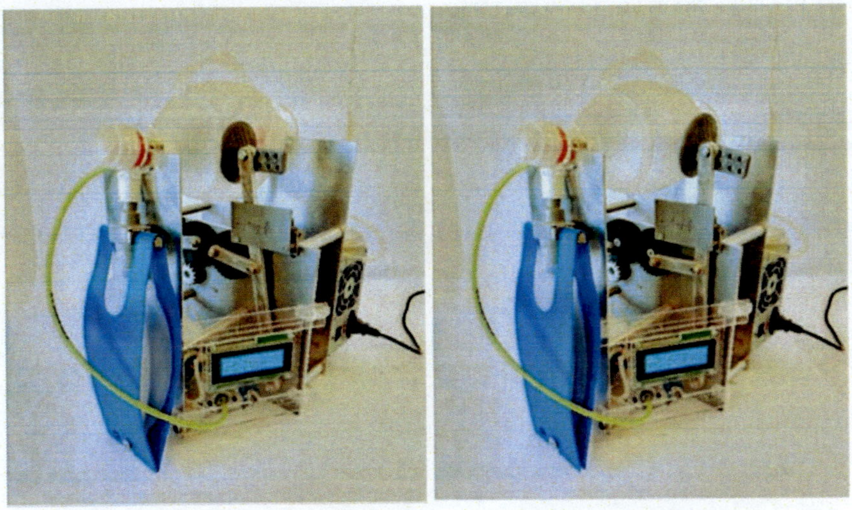

Fig. 11 PCMV prototype: inspiration semi cycle (left), expiration semi cycle (right)

design; model based design and "hardware in the loop" simulations. In this work, a simulation was carried out in Simulink to reproduce the dynamical response of the process variables of the mechanical ventilations system. A transfer functions that reproduces the airflow as a function of input pressure was analyzed to better understand the functioning of the PCMV. This modeling in continuous time was latter converted to discrete time and implemented in LabVIEW connected to the microcontroller in a HiL simulation scheme. Control software functions were verified prior to complete the hardware-software integration of the system prototype. The mechatronic design methodology was reinterpreted and defined in seven stages with clearly identified inputs and outputs for each system design and system integration stages. The simplified methodology was found to be useful for the development of a system prototype of Technology Readiness Level (TRL) of 7, it was possible the development within eight weeks of a system prototype validated and verified for a certain target environment. The presented methodology for mechatronic design will be used to teach mechatronic design and as a standard methodology for future development of mechatronic systems.

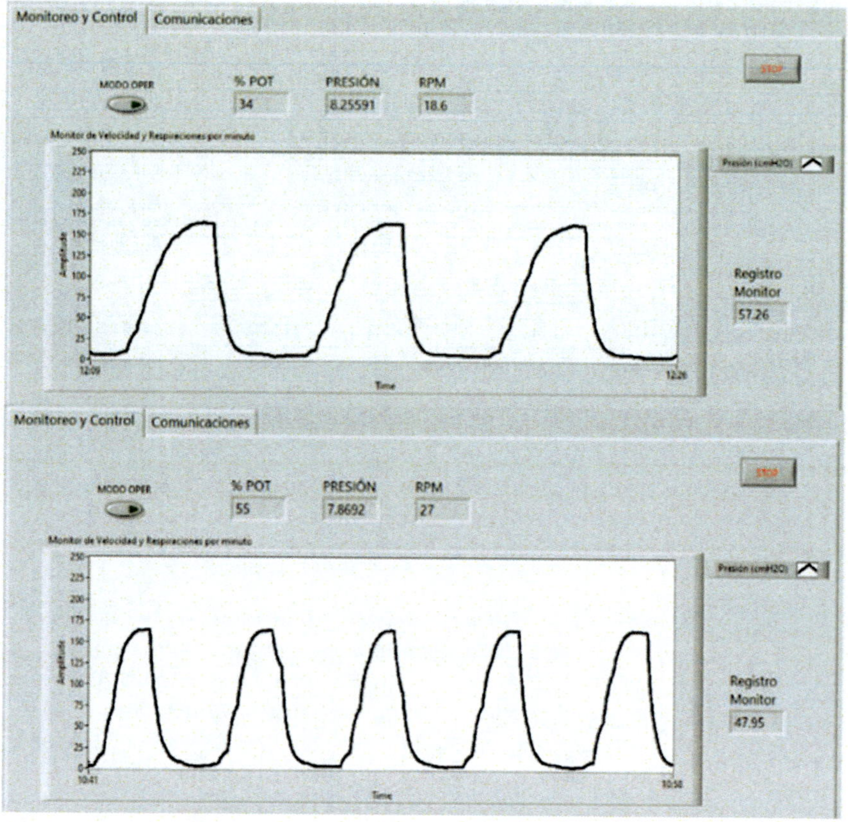

Fig. 12 PCMV pressure measurement at 18.6 rpm (upper) and 27 rpm (lower)

References

1. Isermann R (2008) Mechatronic systems: innovative products with embedded control. Control Eng Pract 16(1):14–29
2. Bishop RH (ed) (2002) The mechatronics handbook, 1st edn. CRC Press
3. Shetty D, Kolk RA (2011) Mechatronics system design, 2nd edn. Cengage Learning
4. Gorrostieta E, Vargas-Soto E, Zuñiga-Aviles L, Rodriguez-Resendiz J, Tovar-Arriaga S (2015) Mechatronics methodology: 15 years of experience. Ingeniería e Investigación 35(3):107–114
5. Cairo JM (2012) Pilbeam's mechanical ventilation, 5th edn. Mosby Inc., St. Louis, MO
6. Victor MH, Forgiarini LA, Kinjo TM, Amato MBP, Yoneyama T, Tanaka H (2015) Parameter estimation of an artificial respiratory system under mechanical ventilation following a noisy regime. Res Biomed Eng 31(4):343–351
7. Hao L et al (2019) Dynamic characteristics of a mechanical ventilation system with spontaneous breathing. IEEE Access 7:172847–172859. https://doi.org/10.1109/ACCESS.2019.2955075
8. Al-Naggar NQ (2015) Modelling and simulation of pressure, controlled mechanical ventilation system. J Biomed Sci Eng 8:707–716
9. Khoo MCK (2001) Physiological control systems: analysis, simulation, and estimation. IEEE Press Series on Biomedical Engineering, New York, pp 1–319

10. Chatburn RL, Daoud EG (2012) Ventilation. In: Kacmarek RM, Stoller JK, Heuer AH (eds) Egan's fundamentals of respiratory care, 10th edn. Mosby Elsevier, St. Louis, MO, pp 225–249
11. Sassoon CSH (2011) Triggering of the ventilator in patient-ventilator interactions. Respir Care 56(1):39–48
12. Abba A et al (2021) The novel mechanical ventilator Milano for the COVID-19 pandemic. Phys Fluids 33:037122. https://doi.org/10.1063/5.0044445

Model Reduction and Control Design of a Multi-agent Line Formation of Mobile Robots

Adrian-Josue Guel Cortez and Eun-jin Kim

Abstract In this work, a model reduction of a line robotic formation driven by simple *PD*-controllers is presented. The proposed mathematical model describes the control interactions between the agents which permits us to easily design a decentralised control strategy. To select the *PD*-controller gains for each agent, we employ a population-based algorithm that takes into consideration the formation stability analysis. Finally, we discuss the future work and the manner the proposed methodology can be used in more complex robotic scenarios.

Keywords Multi-agent systems · Line formation · Evolutionary algorithms · Control design.

1 Introduction

Control of multi-agent systems (MAS) is currently a trend topic with plenty of different approaches [1, 2]. The study of MAS includes not only physical systems but cyberphysical systems [3], biological neuronal networks [4], big data [5] and social networks [6]. This makes MAS a very multidisciplinary and complicated area of research. One of the main issues, where significant efforts have been made is the design of control techniques. This is because MAS control implies a variety of challenges including: modelling, scalability and communication [7].

MAS control design can be simplified if we obtain simpler and reliable mathematical models. For instance, to find a linear representation of the interactions between agents that also includes the control actions. If every agent is controlled by simple low-order controllers, we may use well known linear system control design techniques (see [8, 9]) or new low order control algorithms (see [10–12]). On the other hand, there are other methodologies that could be explored, for example the use of

A.-J. G. Cortez (✉) · E. Kim
Centre for Fluid and Complex Systems, Coventry University, Priory St, Coventry CV1 5FB, UK
e-mail: guelcortea@uni.coventry.ac.uk

K. L. Flores Rodríguez et al. (eds.), *Recent Trends in Sustainable Engineering*, Lecture Notes in Networks and Systems 297,
https://doi.org/10.1007/978-3-030-82064-0_16

information metrics for guided self-organisation in the MAS formation [13] or the use of a fokker-planck control approach [14, 15].

Considering the given context and inspired by our previous experience with a robotic formation experimental setup (see [16]), in this work, we propose a model reduction of a n-agent line formation driven by simple *PD*-controllers. Our model considers that the agents need to keep a desired distance between each other while being limited to see only the agent in front of them. Besides, the nature of the proposed model allows us to study its stability by means of simple linear systems techniques. To solve the control design problem, we use an evolutionary algorithm to find the set of controller gains that improve the system dynamics. In this regard, a discussion on some simulation results is included. Finally, a set of conclusions which include the future work is given.

2 Mathematical Model

Consider a system of two robots moving in the x-axis interacting within each other through a control action that keeps them apart by a desired distance d. The controller is described as an operator L over the relative error ϵ as shown in the left of Fig. 1.

In our case, the operator L is given by a classical *PD* controller, $L = k_p + k_d \frac{d}{dt}$. Here ϵ corresponds to the relative error between the position q_1 and q_2 of the mobile robots, and F_i represents the control force applied to the i-th robot. Note that we can use a spring and a damper in parallel to describe our *PD*'s control action. Specifically, for a two-robot system, we proceed as follows:

The time-evolution of m_1 and m_2 is governed by

$$m_1\ddot{q}_1 + f_{v,1}\dot{q}_1 = F_1, \tag{1}$$

$$m_2\ddot{q}_2 + f_{v,2}\dot{q}_2 = F_2. \tag{2}$$

Since $F_2 = L\epsilon = L(d - (q_2 - q_1))$ and $F_2 = -F_1$ we have

$$m_1\ddot{q}_1 + f_{v,1}\dot{q}_1 = k_p(-d - (q_1 - q_2)) - k_d(\dot{q}_1 - \dot{q}_2), \tag{3}$$

$$m_2\ddot{q}_2 + f_{v,2}\dot{q}_2 = k_p(d - (q_2 - q_1)) - k_d(\dot{q}_2 - \dot{q}_1). \tag{4}$$

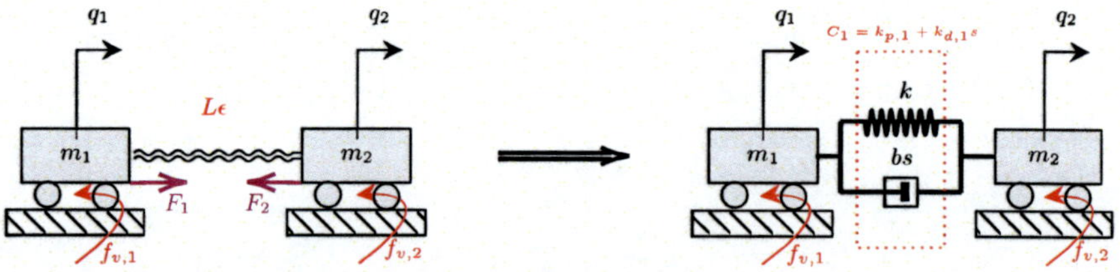

Fig. 1 Interaction of 2 robots in 1D

Equations (3)–(4) give the dynamics of the system shown on the right-hand side of Fig. 1.

In a two robot system, if m_2 corresponds to the robot formation leader and it does not care about the position of m_1, the force $F_2 = 0$.

2.1 1D Model for n Robots

We can generalise the previous procedure to the case of n mobile robots. If the robots can only spot the robot in front, they form a line moving in one dimension and are controlled by a classical *PD* controller as shown in Fig. 2. Then, considering the change of variable $q_{i,1} = q_i$ and $q_{i,2} = \dot{q}_i$, the operator $L_i = k_{p,i} + k_{d,i}\frac{d}{dt}$ and $\epsilon_i = -d_i - (q_i - q_{i+1})$, the mathematical model describing the interaction between the n robots is given by

$$\dot{\mathbf{q}} = \mathbf{A}\mathbf{q} + \mathbf{B}\mathbf{u}, \tag{5}$$

where

$$\mathbf{A} = \begin{bmatrix} \mathbf{H} & \dots & \mathbf{0} \\ \vdots & \ddots & \vdots \\ \mathbf{0} & \dots & \mathbf{O} \end{bmatrix}, \mathbf{H} = \begin{bmatrix} 0 & 1 & 0 & 0 & 0 & 0 \\ -\frac{k_{p,1}}{m_1} & -\frac{k_{d,1}+f_{v,1}}{m_1} & \frac{k_{p,1}}{m_1} & \frac{k_{d,1}}{m_1} & 0 & 0 \\ 0 & 0 & 0 & 1 & 0 & 0 \\ 0 & 0 & -\frac{k_{p,2}}{m_2} & -\frac{k_{d,2}+f_{v,2}}{m_2} & \frac{k_{p,2}}{m_2} & \frac{k_{d,2}}{m_2} \\ 0 & 0 & 0 & 0 & 0 & 1 \\ 0 & 0 & 0 & 0 & -\frac{k_{p,3}}{m_3} & -\frac{k_{d,3}+f_{v,3}}{m_3} \end{bmatrix},$$

$$\mathbf{O} = \begin{bmatrix} 0 & 1 & 0 & 0 \\ -\frac{k_{p,n-1}}{m_n-1} & -\frac{k_{d,n-1}+f_{v,n-1}}{m_n-1} & \frac{k_{p,n-1}}{m_n-1} & \frac{k_{d,n-1}}{m_n-1} \\ 0 & 0 & 0 & 1 \\ 0 & 0 & 0 & -\frac{f_{v,n}}{m_n} \end{bmatrix},$$

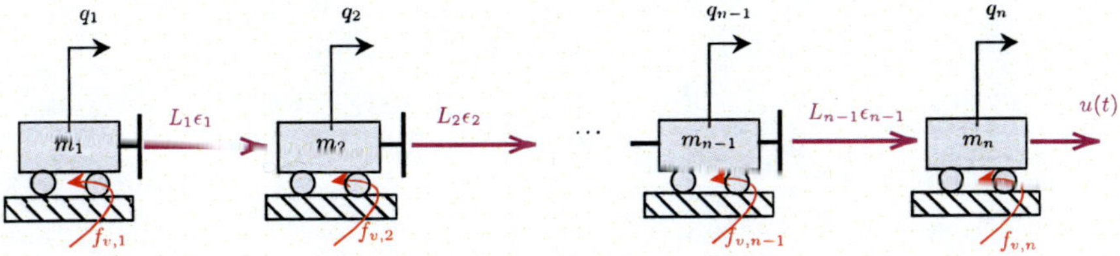

Fig. 2 Interaction of n robots in 1D

$$\mathbf{q} = \left[q_{1,1} \; q_{1,2} \; q_{2,1} \; q_{2,2} \; q_{3,1} \; q_{3,2} \; \cdots \; q_{n-1,1} \; q_{n-1,2} \; q_{n,1} \; q_{n,2} \right]^T,$$

$$\mathbf{Bu} = \left[0 \; -\frac{k_{p,1}d_1}{m_1} 0 \; -\frac{k_{p,2}d_2}{m_2} \; 0 \; -\frac{k_{p,3}d_3}{m_3} \; \cdots \; 0 \; -\frac{k_{p,n-1}d_{n-1}}{m_{n-1}} 0 \; u(t) \right]^T. \quad (6)$$

Or in the form of the recurrence equations

$$\begin{aligned}
\dot{q}_{i,1} &= q_{i,2} \\
\dot{q}_{i,2} &= \frac{1}{m_i} \left(-k_{p,i} q_{i,1} - (f_{v,i} + k_{d,i}) q_{i,2} + k_{p,i} q_{i+1,1} + k_{d,i} q_{i+1,2} - d_i k_{p,i} \right), \\
\dot{q}_{n,1} &= q_{n,2} \\
\dot{q}_{n,2} &= \frac{1}{m_n} \left(-f_{v,n} q_{n,2} + u(t) \right).
\end{aligned} \quad (7)$$

where $i = 1, 2, \ldots, n-1$.

Remark 1 Equation (5) models the interactions between the robots but not fully describes the dynamics of each robot individually. We consider that Eq. (5) is a good approximation to the dynamics of the robotic formation when the robots move with low velocities. This is a feasible assumption in some practical scenarios.

3 Stability Analysis

Since our process is a simple linear model, we can use any linear systems stability analysis methodology. For instance, by defining the system's output as $\mathbf{Y} = \mathbf{Cq} + \mathbf{D}$, we can use the expression $G(s) = \mathbf{C}(s\mathbf{I} - \mathbf{A})^{-1}\mathbf{B} + \mathbf{D}$ to find a transfer function to do the analysis in the Laplace domain. To clarify this idea, let us consider the case where we have three mobile robots. Then, for $n = 3$ in Eq. (5) we have

$$\dot{\mathbf{q}} = \begin{bmatrix} 0 & 1 & 0 & 0 & 0 & 0 \\ -\frac{k_{p,1}}{m_1} & -\frac{k_{d,1}+f_{v,1}}{m_1} & \frac{k_{p,1}}{m_1} & \frac{k_{d,1}}{m_1} & 0 & 0 \\ 0 & 0 & 0 & 1 & 0 & 0 \\ 0 & 0 & -\frac{k_{p,2}}{m_2} & -\frac{k_{d,2}+f_{v,2}}{m_2} & \frac{k_{p,2}}{m_2} & \frac{k_{d,2}}{m_2} \\ 0 & 0 & 0 & 0 & 0 & 1 \\ 0 & 0 & 0 & 0 & 0 & -\frac{f_{v,3}}{m3} \end{bmatrix} \begin{bmatrix} q_{1,1} \\ q_{1,2} \\ q_{2,1} \\ q_{2,2} \\ q_{3,1} \\ q_{3,2} \end{bmatrix}$$
$$+ \begin{bmatrix} 0 & 0 & 0 & 0 & 0 & 0 \\ 0 & -\frac{k_{p,1}}{m_1} & 0 & 0 & 0 & 0 \\ 0 & 0 & 0 & 0 & 0 & 0 \\ 0 & 0 & 0 & -\frac{k_{p,2}}{m_2} & 0 & 0 \\ 0 & 0 & 0 & 0 & 0 & 0 \\ 0 & 0 & 0 & 0 & 0 & 1 \end{bmatrix} \begin{bmatrix} 0 \\ d_1 \\ 0 \\ d_2 \\ 0 \\ u(t) \end{bmatrix}. \quad (8)$$

Specifically, let us assume that we can write the system's output as

$$\mathbf{Y} = \begin{bmatrix} y_1 \\ y_2 \\ y_3 \end{bmatrix} = \begin{bmatrix} 1\,0\,0\,0\,0\,0 \\ 0\,0\,1\,0\,0\,0 \\ 0\,0\,0\,0\,1\,0 \end{bmatrix} \begin{bmatrix} q_{1,1} \\ q_{1,2} \\ q_{2,1} \\ q_{2,2} \\ q_{3,1} \\ q_{3,2} \end{bmatrix}, \tag{9}$$

i.e. by measuring the position of every agent in the system.

Then, we can obtain the MIMO transfer function

$$G(s) = \begin{bmatrix} 0 & G_{12} & 0 & G_{14} & 0 & G_{16} \\ 0 & 0 & 0 & G_{24} & 0 & G_{26} \\ 0 & 0 & 0 & 0 & 0 & G_{36} \end{bmatrix}, \tag{10}$$

where

$$\begin{aligned} G_{12} &= -\frac{k_{p,1}}{s(f_{v,1} + k_{d,1} + m_1 s) + k_{p,1}}, \\ G_{14} &= -\frac{k_{p,2}(k_{d,1}s + k_{p,1})}{(s(f_{v,1} + k_{d,1} + m_1 s) + k_{p,1})(s(f_{v,2} + k_{d,2} + m_2 s) + k_{p,2})}, \\ G_{16} &= \frac{m_3(k_{d,1}s + k_{p,1})(k_{d,2}s + k_{p,2})}{s(f_{v,3} + m_3 s)(s(f_{v,1} + k_{d,1} + m_1 s) + k_{p,1})(s(f_{v,2} + k_{d,2} + m_2 s) + k_{p,2})}, \\ G_{24} &= -\frac{k_{p,2}}{s(f_{v,2} + k_{d,2} + m_2 s) + k_{p,2}}, \\ G_{26} &= \frac{m_3(k_{d,2}s + k_{p,2})}{s(f_{v,3} + m_3 s)(s(f_{v,2} + k_{d,2} + m_2 s) + k_{p,2})}, \\ G_{36} &= \frac{m_3}{f_{v,3}s + m_3 s^2}. \end{aligned} \tag{11}$$

As we can see from Eq. (11), the stability of our process can be studied by the characteristic polynomial of G_{16} which is given by

$$P(s) = s(f_{v,3} + m_3 s)(s(f_{v,1} + k_{d,1} + m_1 s) + k_{p,1})(s(f_{v,2} + k_{d,2} + m_2 s) + k_{p,2}). \tag{12}$$

Note that the polynomial (12) contains the roots of the root of the transfer functions' characteristic polynomials.

Since all the parameters in the system are positive, by taking $k_{p,1}, k_{d,1}, k_{p,2}, k_{d,2} > 0$ we would make the system stable (Hurwitz polynomial). We could further add time delays to the process to use more elaborated stability analysis methodologies (for instance, see [12]).

It is important to notice that a selection of gains that stabilises the process does not guarantee a proper behaviour. For instance, we consider the pair of simulations shown in Fig. 3. Here, even though, both simulations use stabilising gains, one of them shows a crashing between the robots. In such robotic formations, the more robots in the process the more susceptible the system is to transmit oscillations.

Therefore, to design proper controller gains, we must select the transfer function from Eq. (11) that provides us with the most useful transient dynamical information. For the case of three robots, this corresponds to $G_{14}(s)$ because it relates the distance d_2 with the mobile's position q_1. In other words, $G_{14}(s)$ relates the dynamics of the mobiles next and farthest to the leader. Let us recall that every agent can only see who is in front and not who is behind of itself.

4 Algorithm for the Controller's Design

Designing a good controller depends on the goal to be achieved. Thus, our problem can be described as an optimisation problem in a parameters' hyperspace (the space of the controllers gains). We can formulate the optimisation problem as (Fig. 4).

$$\min_{\mathbf{k}} [\mathcal{J}(t)] \text{ s.t. } \mathbf{k} \in \mathbf{S} \tag{13}$$

where $\mathbf{k} \in \mathbb{R}^n$ is the set of control parameters inside the stability region $\mathbf{S}$ and $\mathcal{J}$ is our cost function. In this work we use the Root-Mean-Square Deviation (RMSD) given by:

$$\mathcal{J} = \sqrt{\frac{\sum_{t=1}^{T}(u_t - \hat{y}_t)^2}{T}}. \tag{14}$$

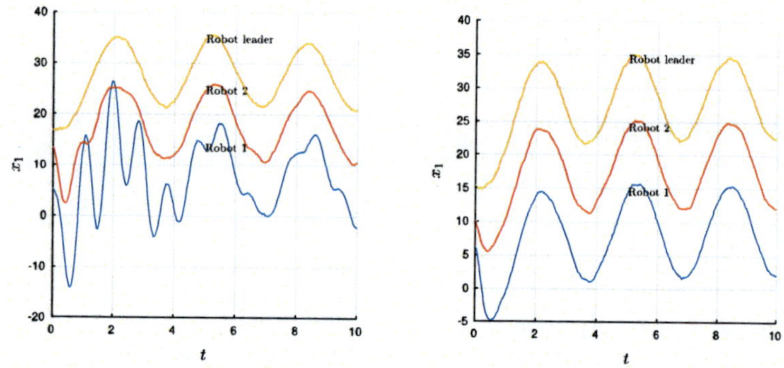

(a) Undesired behaviour. There exist crashing between the robots. The values $k_p = 8$, $k_d = 1$ are used in every robot.

(b) Desired behaviour. There is no crashing between the robots. The values $k_p = 50$, $k_d = 10$ are used in every robot.

Fig. 3 System of 3 robots. $u(t) = 100\sin(t)$, all masses $m_i = 1, f_{v,i} \in (0, 2)$ and all $d_i = 10$

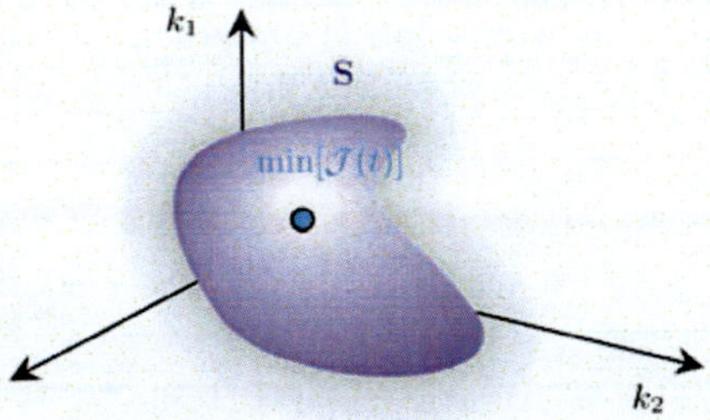

Fig. 4 Diagram of the optimisation problem. Finding the vector **k** inside the stability region $\mathcal{S}$, that minimise the RMSD $\mathcal{J}$ in a 3rd order problem

Here, $\hat{y}_t$ is the output of the transfer function with the characteristic polynomial $P(s)$ using a set of control parameters **k** with an input equal to the desired response u_t. T is the total number of iterations.

4.1 Solution to the Optimisation Problem

Due to the complexity of the optimisation problem formulated by Eq. (13), in this section, an evolutionary algorithm is proposed as an starting point. We describe it as a pseudocode in Algorithm 1.

Some important features of this algorithm are the way in which we compute the score of each element in the population and the way in which we create the new population. For the latter, we use a Blend crossover operator (BLX) [17]. The fitness/score is computed by using the exp(−RMSD) in order to normalise the RMSD while penalising big RMSDs.

5 Results

In this section, we discuss results of some simulations of system (5) for the case $n = 3$ subject to a set of controller gains **k** designed by means of Algorithm 1. For the present simulations, we have computed the RMSD of the time response of $G_{14}(s)$ using two different reference signals $u(t) = 1$ (regulation) and $u(t) = 30\sin(2t)$ (tracking) in Algorithm 1. Besides, we have used all $m_i = 1$ and $f_{v,i} = 0.5$ while keeping all agents desired distance between each other $d_i = 10$ in our simulations. Additionally, we have kept the maximum number of iterations of Algorithm 1 equal to 100. The simulation results are given in Figs. 5, 6, 7 and 8.

First, the results of the case $u(t) = 1$ are shown in Figs. 5 and 6. Here, Fig. 5 presents the case when the elements k_i of the controller gains vector **k** are in the range $0 < k_i < 500$, this large range of values allow us to have a faster step response of $G_{14}(s)$ as it is depicted in Fig. 5a (fast an overdamped response). Let us recall that by using the RMSD as a cost function, we force Algorithm 1 to find the set

Algorithm 1: Control design algorithm.

Data: Consider the initial population $\mathcal{P}_0 := \{\mathbf{k}_1, \ldots, \mathbf{k}_N | \mathbf{k}_i \in \mathcal{S} \forall i = 1, \ldots, N\}$ where $\mathcal{S} \subset \mathbb{R}^n$ is the stability region of the parameter space $\mathbf{k}$, the input $u(t)$ and the total simulation time t_f.

Result: The value of the gain vector $\mathbf{k}_{\hat{i}}$ for which $\mathcal{J}$ is minimum over the time $t \in (0, t_f)$ for the input $u(t)$.

/* Score the initial population $\mathcal{P}_0$ */

$\mathbf{F} := [\mathcal{F}_1, \ldots, \mathcal{F}_N]^T$ // Vector with computed fitness of each element in the population.

$\mathbf{E} := [E_1, \ldots, E_N]^T$ // Vector with computed RMSDs of each element in the population.

/* Compute the fitness value. */

for $i \leftarrow 1$ **to** N **do**

  $E_i = \mathcal{J}(\mathbf{k}_i \in \mathcal{P}_0, u, t_f)$

  $\mathcal{F}_i = e^{-E_i}$

end

$f_b = \min(|1 - \mathbf{F}|)$ // Find the best score f_b in the initial population and compare it with a desired goal d_g.

/* Repeat the process until the desired goal is accomplish */

while $f_b > d_g$ **do**

  $\mathcal{P}_N = \mathrm{BLX}(\mathcal{P}_0)$ // Create a new population $\mathcal{P}_N$ using the BLX algorithm on the old population $\mathcal{P}_0$.

  for $i \leftarrow 1$ **to** N **do**

    $E_i = \mathcal{J}(\mathbf{k}_i \in \mathcal{P}_N, u, t_f)$

    $\mathcal{F}_i = e^{-E_i}$

  end

  $\mathcal{P}_0 = \mathcal{P}_N$ // The new population replaces the old one.

  $f_b = \min(|1 - \mathbf{F}|)$

end

$\hat{i} = \mathrm{find}(|1 - \mathbf{F}| = f_b)$ // Find the position $\hat{i}$ of $\mathbf{k}_{\hat{i}} \in \mathcal{P}_N$ with the best score.

return $\mathbf{k}_{\hat{i}} \in \mathcal{P}_N$

of gains $\mathbf{k}$ that minimise the difference between $u(t)$ and $y(t)$ for a given time. In our simulations, we have used a final simulation time $t_f = 1$ to get a fast response. Using a larger t_f would make our algorithm computationally more expensive. In addition, Fig. 5b shows that every agent lasts at least 2 seconds to reach to the desired separation between each other. This agrees with the step response of $G_{14}(s)$ subject to the designed controller gains $\mathbf{k}$ (see Fig. 5a). On the other hand, Fig. 6 represents a similar scenario although here we have diminished the range at which ever element k_i of the set of controller gains $\mathbf{k}$ is generated. This gives us a smooth but slower step response as indicated in both the step response of $G_{14}(s)$ (see Fig. 6a) and the time response of the robotic formation (see Fig. 6b).

For the case when $u(t) = 30\sin(2t)$, we have again implemented our design algorithm by changing the range of possible values that every element k_i of our controller $\mathbf{k}$ can take. Figure 7 shows the case when the elements k_i of the controller gains vector $\mathbf{k}$ are in the range $0 < k_i < 50$. This gives a set of gains that does not completely

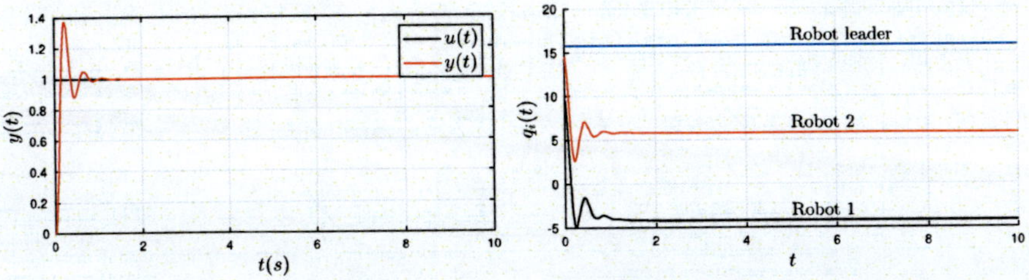

(a) $G_{14}(s)$ step response. (b) Formation behaviour.

Fig. 5 System of 3 robots. $u(t) = 0$, all $m_i = 1$, $f_{v,i} = 0.5$, $d_i = 10$ and the elements k_i of the controller gains vector $\mathbf{k}$ are in the range $0 < k_i < 500$. We iterate Algorithm 1 one hundred times. The final value of $\mathbf{k} = [k_{p,1}, k_{d,1}, k_{p,2}, k_{d,2}]^T = [291.2359, 103.4591, 229.6364, 9.0647]^T$

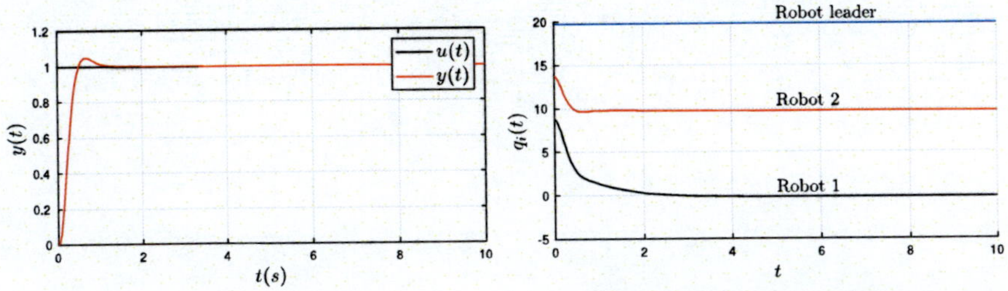

(a) $G_{14}(s)$ step response. (b) Formation behaviour.

Fig. 6 System of 3 robots. $u(t) = 0$, all $m_i = 1$, $f_{v,i} = 0.5$, $d_i = 10$ and the elements k_i of the controller gains vector $\mathbf{k}$ are in the range $0 < k_i < 50$. We iterate Algorithm 1 one hundred times. The final value of $\mathbf{k} = [k_{p,1}, k_{d,1}, k_{p,2}, k_{d,2}]^T = [37.8257, 33.5179, 48.2762, 9.5010]^T$

reduce the value of RMSD, as it is shown in the time response of $G_{14}(s)$ (see Fig. 7a). Nonetheless, for this set of values the robotic formation shows a good performance (see Fig. 7b). On the contrary, Fig. 8, shows the case when the elements k_i are in the range $0 < k_i < 500$. This allows us to obtain a smaller RMSD value, as we can see from the time response in Fig. 8a, but it also adds small oscillations at the beginning of the robotic formation simulation (see Fig. 8b).

6 Conclusions

This work presents a linear model reduction for a line formation of mobile robots driven by classical *PD*-controllers. This model permits us to easily design stabilising controllers for each agent in the formation. The proposed model can be generalised to study systems in higher dimensions by considering the particles in a plane. We have also analysed the use of an evolutionary algorithm to find the best set of control

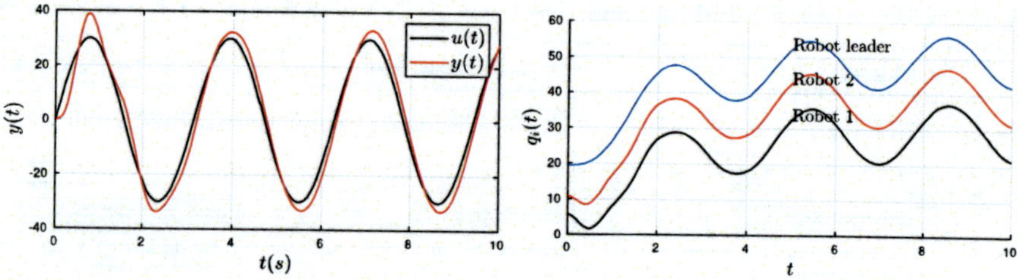

(a) $G_{14}(s)$ time response. (b) Formation behaviour.

Fig. 7 System of 3 robots. $u(t) = 30\sin(2t)$, all $m_i = 1, f_{v,i} = 0.5$, $d_i = 10$ and the elements k_i of the controller gains vector **k** are in the range $0 < k_i < 50$. We iterate Algorithm 1 one hundred times. The value of $\mathbf{k} = [k_{p,1}, k_{d,1}, k_{p,2}, k_{d,2}]^T = [31.4295, 32.2729, 45.6248, 0.7629]^T$

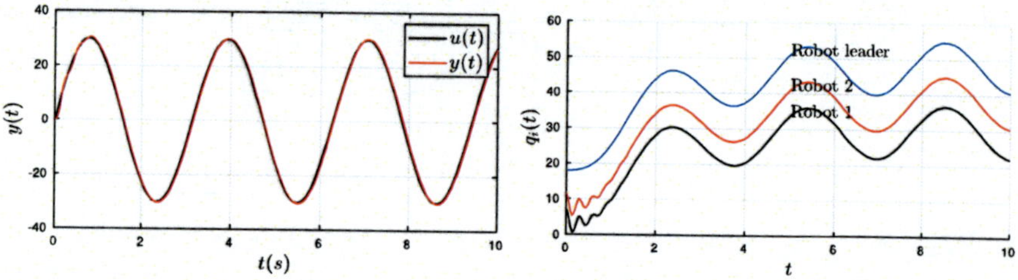

(a) $G_{14}(s)$ time response. (b) Formation behaviour.

Fig. 8 System of 3 robots. $u(t) = 30\sin(2t)$, all $m_i = 1, f_{v,i} = 0.5$, $d_i = 10$ and the elements k_i of the controller gains vector **k** are in the range $0 < k_i < 500$. We iterate Algorithm 1 one hundred times. The value of $\mathbf{k} = [k_{p,1}, k_{d,1}, k_{p,2}, k_{d,2}]^T = [28.0426, 213.5836, 440.8379, 5.3353]^T$

parameters that improves the network performance. The results of our algorithm are discussed together with some simulations.

Future work includes exploring several novel and well known linear control design techniques (for instance, see [9, 10]), the generalisation of the model to a multi-dimensional framework and the improvement of the optimisation method for the controller's design.

References

1. Knorn S, Chen Z, Middleton RH (2015) Overview: collective control of multiagent systems. IEEE Trans Control Netw Syst 3(4):334–347
2. Bussmann S, Jennings NR, Wooldridge M (2013) Multiagent systems for manufacturing control: a design methodology. Springer Science & Business Media
3. Zhang D, Feng G, Shi Y, Srinivasan D (2021) Physical safety and cyber security analysis of multi-agent systems: a survey of recent advances. IEEE/CAA J Automatica Sinica 8(2):319–333

4. Ma J, Tang J (2015) A review for dynamics of collective behaviors of network of neurons. SCIENCE CHINA Technol Sci 58(12):2038–2045
5. Giannakis M, Louis M (2016) A multi-agent based system with big data processing for enhanced supply chain agility. J Enterpr Inf Manage
6. Rodriguez MA, Steinbock DJ, Watkins JH, Gershenson C, Bollen J, Grey V, Degraf B (2007) Smartocracy: Social networks for collective decision making. In: 2007 40th annual Hawaii international conference on system sciences (HICSS'07). IEEE, pp 90–90
7. Oh KK, Park MC, Ahn HS (2015) A survey of multi-agent formation control. Automatica 53:424–440
8. Åström KJ, Hägglund T (2001) The future of pid control. Control Eng Pract 9(11):1163–1175
9. Méndez-Barrios C, Niculescu SI, Morarescu CI, Gu K (2008) On the fragility of pi controllers for time-delay SISO systems. In: 2008 16th Mediterranean conference on control and automation. IEEE, pp 529–534
10. Guel-Cortez AJ, Méndez-Barrios CF, Kim EJ, Sen M (2021) Fractional-order controllers for irrational systems. IET Control Theor Appl
11. Guel-Cortez AJ, Méndez-Barrios CF, González-Galván EJ, Mejía-Rodríguez G, Félix L (2019) Geometrical design of fractional pdμ controllers for linear time-invariant fractional-order systems with time delay. Proce Inst Mech Eng Part I J Syst Control Eng 233(7):815–829
12. Hernández-Díez JE, Méndez-Barrios CF, Mondié S, Niculescu SI, González-Galván EJ (2018) Proportional-delayed controllers design for lti-systems: a geometric approach. Int J Control 91(4):907–925
13. Guel-Cortez AJ, Kim EJ (2020) Information length analysis of linear autonomous stochastic processes. Entropy 22(11):1265
14. Annunziato M, Borzì A (2013) A fokker-planck control framework for multidimensional stochastic processes. J Comput Appl Math 237(1):487–507
15. Mwaffo V, DeLellis P, Humbert JS (2021) Formation control of stochastic multivehicle systems. IEEE Trans Control Syst Technol
16. Ramos-Á vila D, Rodriguez C, Hernández-Carrillo J, Guel-Cortez AJ, Sen M, Méndez-Barrios CF, González-Galván E, Goodwine B (2019) Experiments with PD-controlled robots in ring formation. In: XXI Congreso Mexicano de Robótica - COMRob
17. Abido MA (2003) A novel multiobjective evolutionary algorithm for environmental/economic power dispatch. Electr Power Syst Res 65(1):71–81

Design and Modeling of an Elastic Inflatable Actuator to Achieve Single and Multiple Motions Through One Channel

N. Cruz-Santos, D. Martinez-Sanchez, M. Ruiz-Torres, X. Y. Sandoval-Castro, and E. Castillo-Castaneda

Abstract This paper presents the design and modeling of an elastic inflatable actuator, EIA, to generate single and multiple motions through only one channel. The morphology proposed is focused on an external and internal design based on crests with amplitude and length defined. The EIA geometry change in magnitude and direction when pressure or vacuum is applied, generating the multiple motions expanding-contracting. Besides, by adding a thicker layer to a 180° or 120° section in the EIA, a bending motion is generated in the direction where the greatest thickness of the material is located. We describe a detailed design of the EIA and the modeling by FEM analysis for both single and multiple motions. In addition, we introduce a soft linear manipulator by using both single and multiple-motions EIAs, we design a soft gripper with 3 bending motion EIAs; the soft linear manipulator is composed of the multiple motions expanding-contracting EIA connected to the gripper. The manipulator needs only two controlled pressure inputs, the first one generates a linear displacement along its transverse axis and the second one controls the soft gripper closing-opening.

Keywords Soft actuator · Elastomers · Bidirectional movement · Positive pressure · Negative pressure.

1 Introduction

Each time is less the frontier between the robotic devices and humans, nowadays, the soft structures present a friendly and creative way, the solution to tasks that requires the manipulation and grip of fragile objects, amorphous, farm products, and food

N. Cruz-Santos · M. Ruiz-Torres · E. Castillo-Castaneda
Instituto Politecnico Nacional, 76090 Santiago de Queretaro, Qro, Mexico
e-mail: ncruzs1800@alumno.ipn.mx

D. Martinez-Sanchez (✉) · X. Y. Sandoval-Castro
CONACYT-IPN, Mexico City, Mexico
e-mail: dmartinezs1500@alumno.ipn.mx

K. L. Flores Rodríguez et al. (eds.), *Recent Trends in Sustainable Engineering*, Lecture Notes in Networks and Systems 297,
https://doi.org/10.1007/978-3-030-82064-0_17

products [1, 2]; moreover, soft structures could be applied in dynamic and unpredictable tasks, such as inspection of natural disasters and dangerous environments [3, 4]; on the other hand, soft structures also could be used in applications which need a safe interaction with humans, e.g., human tissues, cells and medical applications [5, 6]. For this reason, new designs to achieve complex movements are developed every day. Elastic Inflatable Actuators (EIA) are a kind of soft actuators that are used in the above applications. The EIA's are driven by pressurized fluids (pneumatic or hydraulic) and have the following advantages: lightweight, easy of fabrication, low manufacturing cost, high adaptability, achieving large strokes, exhibit distributed force generation due to Pascal's principle, high power-to-weight ratio, safe operation and ability to provide complex movements with simple control inputs [7–9]. These actuators can achieve four motion, (1) Expanding (axial or radial), (2) Contracting, (3) Twisting, and (4) Bending [7]; to achieve more complex movements it has been proposed to connect two or more morphologies with different motions in a single body [10], which are activate independently through different supply channels; however, the control of this type of actuator has the disadvantage of needing multiple controlled pressure inputs, which increases the cost of the components (pumps, sensors, processor capacity, electronic components, etc.). Due to described above, a soft actuator has been designed to achieve single and multiple motions in the same body, using only one channel; the design is versatile and generates the bending and expanding-contracting motions (without using reinforcement fibers or post-construction materials). The morphology is very complex and was externally and internally carefully designed. We describe the design and the anisotropic program for the expanding-contracting multiple motions EIA, we introduce the anisotropy to achieve bending motion by using the same EIA geometry, varying the thicker layer of one section. We present two soft structures with the same design principle, but each one can achieve different motions. Besides, for each EIA, the FEM modeling results are presented. Finally, we introduce the design and FEM modeling of a soft linear manipulator through both EIAs; by using 3 bending motion EIA, a soft gripper was designed. The soft linear manipulator is composed of the multiple-motions expanding-contracting EIA connected to the soft gripper; the first EIA allows linear displacement along its transverse axis while the gripper allows the amorphous object manipulations. The operating principle of the soft linear manipulator is simple, only two controlled pressure inputs are necessary; the first one controls the expanding-contracting EAI which generates expanding motion along its transverse axis, the second one is used to close and open the soft gripper.

A. **Multiple motions expanding-contracting EIA** To achieve different motion profiles in a soft inflatable actuator, two main parameters must be considered: the morphology programmed in its body and the manufacturing material properties [11]. The EIA morphology is accordion-inspired, it has 10 crests and only one channel; each crest has 21 mm of external diameter with a 10 mm of length; on both sides, left and right, a cover was inserted, and one of them is working as an air supply. Figure 1 shows the expanding-contracting EIA, the design has 112 mm of total length and is completely symmetrical. The expanding motion is

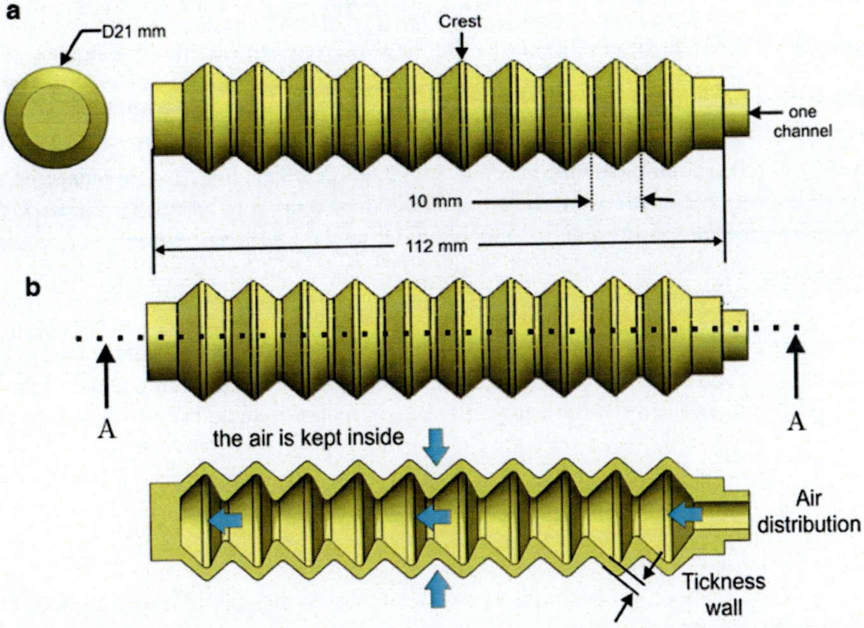

Fig. 1 **a** Design of the multiple motions expanding-contracting EIA. **b** Cross-section view A-A of the expanding-contracting EIA

performed when the air is applied, the pressure begins to flow inside of the EIA, then, each crest is sequentially inflated, which generates a displacement of the actuator body.

Figure 1b shows a cross-section view to appreciate the distribution air. Further, the radial expansion is limited due to the wall thickness in each one of the crests; the wall keeps the air inside the actuator and allows a greater axial displacement of the body.

The external geometry of the EIA is described as sine waves with the same amplitude (A) and length (λ) between them, such as Fig. 2a illustrates.

The key to performing the expanding-contracting motions through a single channel is the internal geometry. The lower crest considers an offset (d), this shape allows the contracting motion when the vacuum is applied. During the contracting motion, the lower crests are aligned in phase with the upper crest. If the structure is manufactured with rubber, a great resistance could be reached due to the rubbers are considered as compressible material; in addition, the rubbers give it the ability to recover its initial shape when the vacuum is removed. Due to the symmetrical design, the mass is homogeneously distributed along the actuator body. Thus, the expanding motion is restricted to only one direction and radial expanding is limited.

B. **Bending EIA** The anisotropy for the bending motion has been programmed in the same internal structure, but now, the external structure has been modified by adding a thicker layer to an actuator's section. Figure 2B-A shows the front view of the expanding-contracting EAI, if the actuator is considered as a cylinder, when we add a greater wall thickness to the half of the body (180°), see Fig. 2B-B, a bending motion is achieved. In the same way, Fig. 2B-C presents the actuator with a greater wall thickness in the third section of its body (120°), the first and the second design have been called alpha and beta, respectively. Figure 2B-D presents a comparison of the transversal-section view expanding-contracting versus bending EIA.

Figure 2c shows a transversal-section view of the alpha actuator's body. The wall thickness of the external geometry was incremented to 1.5 mm, generating bending motion in the direction where is the great material thickness.

2 Modeling

The modeling was performed considering the Ecoflex 00-30 and Ecoflex 00-50 materials for the multiple motions expanding-contracting and bending EIAs, respectively. Figure 3 shows the methodology to achieve the FEM modeling by using specialized software.

FEM modeling of the multiple motions expanding-contracting EIA The modeling for the multiple motions expanding-contracting EIA considers fixing the actuator on the top cover (where the pressure is supplied). Figure 7 presents the FEM results; in the analysis, the expanding-contracting EIA at rest moves 4.30 mm due to gravity (Fig. 7B).

Figure 4A-A shows the axial expanding motion of the EIA, the actuator translates 19.97 mm along the negative Y-axis when 4 kPa of pressure is applied. Besides, Fig. 4A-C presents the contracting motion, the EIA translates 14.85 mm along the positive Y-axis when 1.9 kPa of vacuum is applied.

Figure 8 shows the cross-section view of the EIA. When the expanding motion is performed, the upper and lower crests increase in length (λ) and decrease in amplitude (A) allowing the linear displacement along with the positive Y-axis such as Fig. 4B-A illustrates, further, the morphology restricts the radial expanding. In the same way, when the contracting motion is performed (Fig. 4B-B), the upper and lower crests decrease in length (λ) and increase in amplitude (A), this phenomenon allows decreasing the EIA length and therefore the linear displacement along the negative Y-axis.

FEM modeling of the bending EIA Fig. 5a, b show the FEM results for alpha and beta bending EIAs, respectively. As is displayed in Fig. 5, bending EIA alpha reaches 59.1° while bending EIA beta achieves 52.2° by applying the same pressure (13.9 kPa). The alpha bending EIA has a greater amount of material in a section of the body;

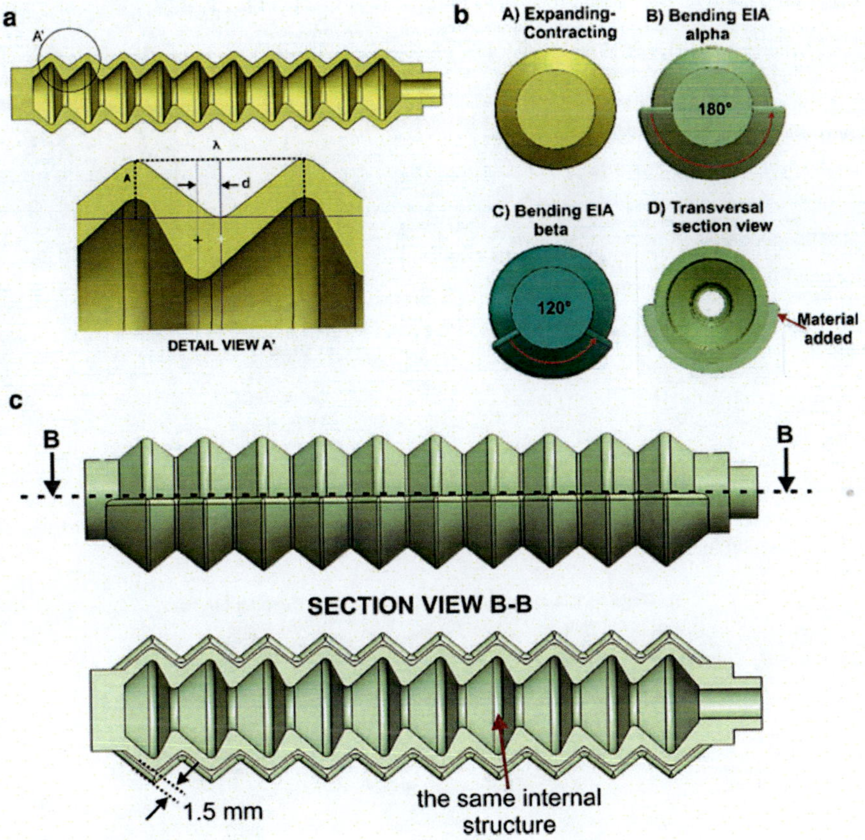

Fig. 2 **a** Detailed view A', external and internal geometry of the expanding-contracting EIA. **b** Front view, (A) expanding-contracting EIA, (B) bending EIA alpha, (C) bending EIA beta, (D) comparison of the transversal-section view (A) versus (B). **c** Transversal-section view B-B, bending alpha EIA

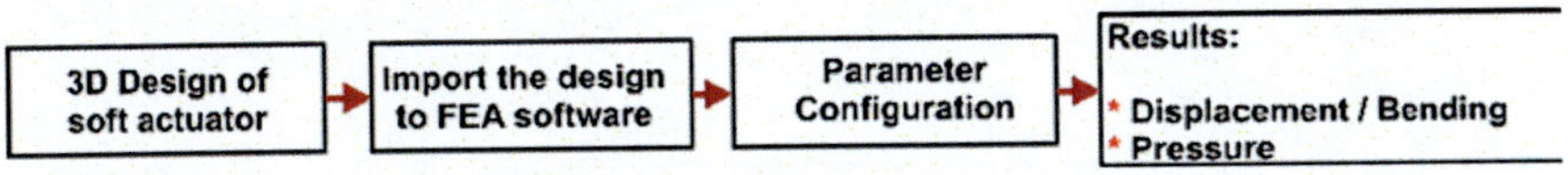

Fig. 3 Methodology to achieve the FEM modeling

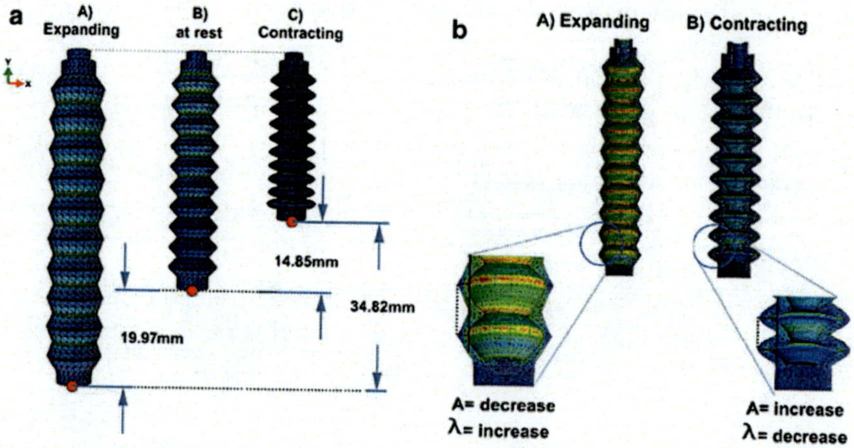

Fig. 4 **a** Multi motions expanding-contracting EIA, FEM modeling. (A) Axial expanding when P = 4 kPa of pressure is applied, (B) EIA at rest, (C) contracting when P = 1.9 kPa of vacuum is applied. **b** Cross-section view of the expanding-contracting EIA during the FEM modeling. (A) Expanding motion, (B) Contracting motion

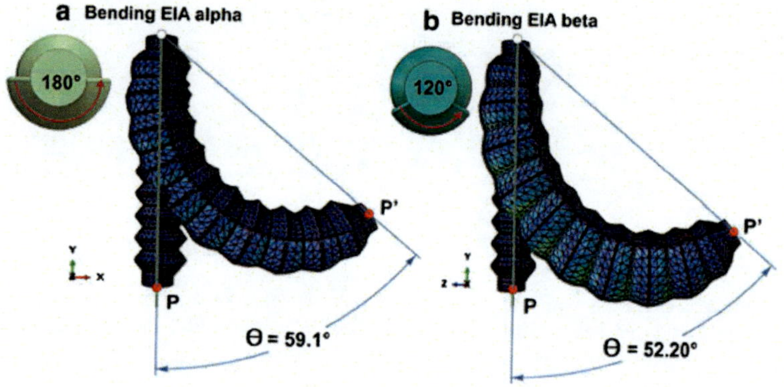

Fig. 5 FEM results for the bending EIA, **a** alpha achieves 59.1°, **b** beta achieves 52.2°

thus, this actuator reaches a higher bending value angle. Otherwise, the beta bending EIA achieves a lower bending value angle since the pressure is not homogeneously distributed into the body; hence, the actuator performs axial and radial expanding motion.

To calculate the bending value angle as a function of the applied value pressure, a node on the below cover was chosen, the coordinates at the rest, P, and final position, P′, were used to compute the bending angle by Eq. (1).

$$\theta = \cos^{-1}(P \cdot P'/|P| \cdot P'|) \tag{1}$$

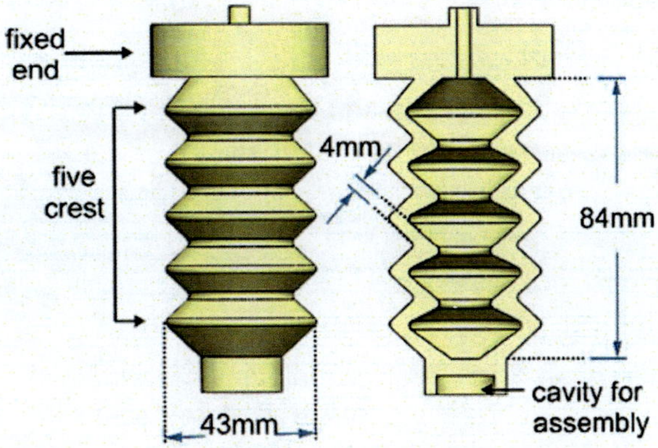

Fig. 6 Design of the extension-contraction actuator for linear manipulator

3 Soft Linear Manipulator

The design of soft structures with complex motions and simple control strategies will facilitate the incursion of soft robotics in daily applications by proposing solutions that adapt to tasks that require delicate and precise handling of amorphous and fragile objects. Due to this, we introduce the design and modeling of a soft linear manipulator, using only the actuators presented in this paper. The manipulator is composed of two elements, an expanding-contracting EIA and a gripper constituted by three bending EIAs.

The expanding-contracting EIA generates a linear displacement along its transverse axis. For the soft manipulation application, we modify the original dimensions of the actuator shown in Fig. 1a to expand its load capacities. Figure 6 presents the geometrical parameters of the expanding-contracting EIA which has five crests of 43 mm diameter, 84 mm of length, and 4 mm of wall thickness.

The assembly of two or more bending soft actuators has been proposed to manipulate food, fragile objects, chemical and pharmaceutical products [6, 12, 13]. A soft gripper based on the bending EIA alpha is presented; Fig. 7a shows three actuators assembled symmetrically separated 120°, this configuration allows distributing the load between the three actuators, the bending EIAs are attached to a hard base, the pressure is supplied to a chamber which guarantees a homogeneous distribution of the air into the three actuators.

Figure 7b shows the FEM modeling of the soft gripper, the gravity acts on the bending EIAs causing a small bending angle at rest, as a consequence, the bottom of the gripper forms an initial diameter (Fig. 7B-A), the coordinates of the bottom side of each actuator (red points) were used to calculate the initial diameter, obtaining 98.83 mm. Figure 7B-B shows the soft gripper when 14 kPa of pressure is applied, allowing the total closing, in this position the bottom side of each actuator is touching between them. This soft gripper provides a large workspace, and the rough surface of the actuators allows greater friction between objects and the gripper.

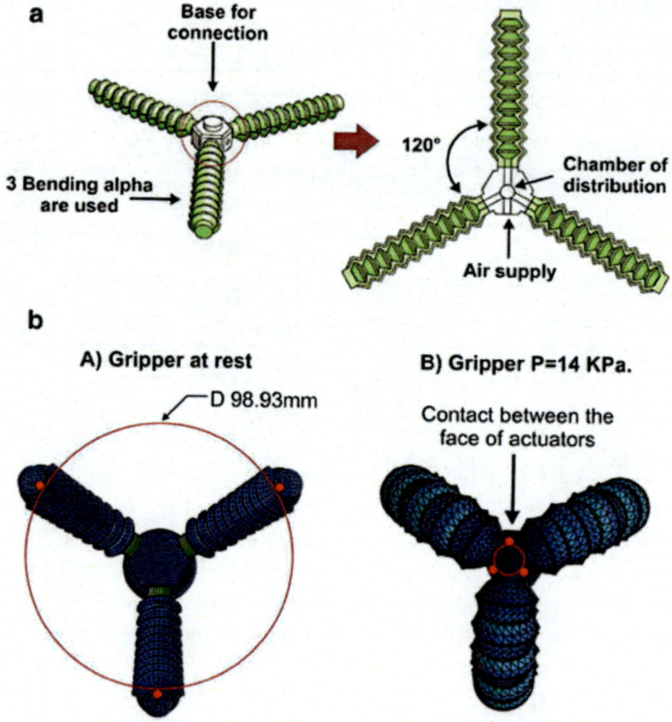

Fig. 7 **a** Soft gripper composed of three bending EIA alpha, separated 120° between them. **b** FEM modeling of the soft gripper

A. **Design and FEM modeling of the soft linear gripper** The soft linear manipulator is a compound of the expanding-contracting EIA and the soft gripper shown in Figs. 6 and 7a, respectively. The expanding-contracting EIA has connected through cavity for assembly to the gripper via the base for connection as is presented in Fig. 8. The expanding-contracting EIA generates a linear displacement along its transverse axis, this manipulator has the only movement along the positive and negative direction of the Z-axis, the bending EIAs of the gripper works as a soft finger that allows the manipulation of an object. The crests around the soft fingers provide a relief that allows better adhesion to the surface of manipulable objects.

The soft linear manipulator presented in Fig. 8 was modeled by using FEM specialized software, considering the material Ecoflex 00-50; to perform the FEM simulations, only two inputs are needed, the first one activates the linear displacement, and the second one activates the opening and closing of the gripper. Figures 9a, b show the FEM results for expanding and contracting motions, respectively.

Figure 9a on left illustrates the soft manipulator at rest, the manipulator shown on the right presents the final deformation when a pressure value of 7 kPa is applied into

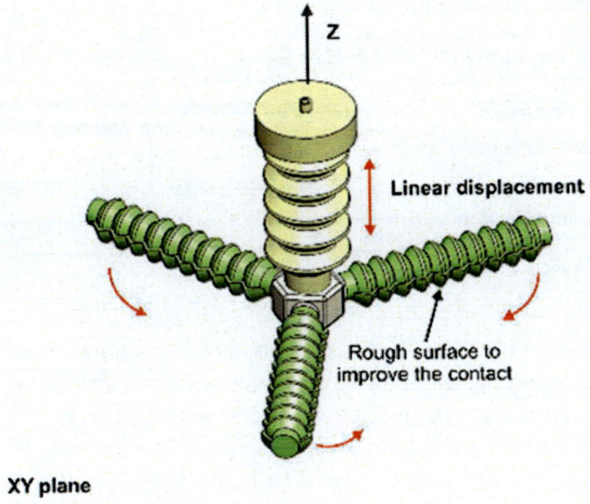

Fig. 8 Design of the soft linear manipulator

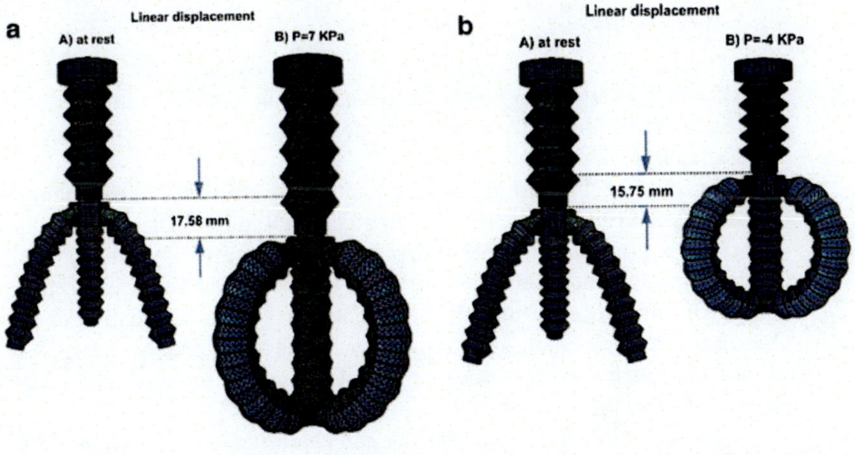

Fig. 9 **a** Soft linear manipulator achieving 17.58 mm and of displacement along the positive direction of the Z-axis. **b** Soft linear manipulator achieving 15.75 mm of displacement along the positive direction of the Z-axis, using negative pressure

the expanding-contracting EIA, achieving a linear displacement of 17.58 mm along the negative direction of the Z-axis, the closed gripper is displayed.

Figure 9b on the left illustrates the soft manipulator at rest, on the right the manipulator is presented when a vacuum of 4 kPa is applied into the expanding-contracting EIA, performing a linear displacement of 15.75 mm along the positive direction of the Z-axis.

It was introduced an EIA with the capability to generate the single and multiple motion profiles of bending and expanding contracting through a single channel.

These motion profiles were programmed in the internal and external structure of the soft actuator, in which geometry is novel and complex.

The EIA morphology is accordion-inspired, the external geometry of the EIA is described as sine waves with the same amplitude and length, the internal geometry considers the same principle, but the below crests are out of phase between them. When the pressure or vacuum is applied to the actuator, the crests change in amplitude and length, generating the multiple motions of expanding-contracting along its transverse axis; obtaining a wide range of linear displacement.

To achieve the single motion of bending, it added a thicker layer to a section on the external geometry of the EIA previously described.

It was presented a FEM modeling of the single and multiple motions EIA. For the expanding-contracting EIA, only one controlled input was considered, by applying pressure, an expanding motion of 19.97 mm is achieved, but applying vacuum, a contracting motion of 14.85 mm is generated. The bending EIA generates angles of 59.1° and 52.2°, for the variants of 180° and 120°, respectively.

4 Conclusions

It is introduced a manipulation application by using only the actuators described in this paper; the bending EIA was used to design a soft gripper compound of three of these actuators.

It is presented the design and FEM modeling of a soft linear manipulator by connecting the gripper with the expanding-contracting EIA. The manipulator requires only two inputs to control it, one of them allows the control of the linear displacement (expanding and contracting motions), and the other one controls the opening and closing of the gripper. This soft manipulator is a novel alternative to manipulate objects with different shapes and dimensions since the gripper has a large workspace; further, the rough surface of the gripper allows improving the contact between the object and the fingers.

As future work, will be manufacture and characterize the soft linear manipulator to determine its load capabilities; an experimental evaluation will be performed, considering different amorphous objects with distinct loads.

References

1. Shintake J, Rosset S, Schubert B, Floreano D (2016) Versatile soft grippers with intrinsic electroadhesion based on multifunctional polymer actuators. Adv Mater 28(2):231–238
2. Martinez RV, Branch JL, Fish CR (2013) Robotic tentacles with three-dimensional mobility based on flexible elastomers. Adv Mater 2:205–212
3. Ishikawa R, Tomita T, Yamada Y, Nakamura T (2016) Development of a peristaltic crawling robot for long-distance complex line sewer pipe inspections. In: 2016 IEEE international con-

ference on advanced intelligent mechatronics (AIM), pp 413–418 (2016). http://orcid.org/10.10007/1234567890
4. Donatelli CM, Serlin ZT, Echols-Jones P (2017) Soft foam robot with caterpillar-inspired gait regimes for terrestrial locomotion. In: IEEE/RSJ international conference on intelligent robots and systems (IROS), pp. 476–481. http://orcid.org/10.1109/IROS.2017.8202196
5. Ranzani T, Gerboni G, Cianchetti M (2015) A bioinspired soft manipulator for minimally invasive surgery. Bioinspir Biomim 10:
6. Do TN, Phan H, Nguyen T-Q (2018) Soft electromagnetic electromagnetic actuators: miniature soft electromagnetic actuators for robotic applications. Adv Func Mater 18(28):1870116
7. Gorissen B, Reynaerts D, Konishi S (2017) Elastic inflatable actuators for soft robotic applications. Adv Mater 28:1–14
8. Agarwal G, Besuchet N, Audergon B (2016) Stretchable materials for robust soft actuators towards assistive wearable devices. Sci Rep 6(1):1–8
9. Robertson M, Sadeghi H, Florezy JM, Paik J (2017) Soft pneumatic actuator fascicles for high force and reliability, soft robotics, pp 23–32
10. Connolly F, Polygerinos P, Walsh CJ (2015) Mechanical programming of soft actuators by varying fiber angle. Soft Rob 2(1):26–32
11. Kim S, Laschiy C, Trimmer B (2013) Soft robotics: a bioinspired evolution, trends in biotechnology, pp 287–288
12. Robertson M, Sadeghi H, Florez JM (2017) Soft pneumatic actuator fascicles for high force and reliability, soft robotics, pp 23–32
13. Rateni G, Cianchetti M, Ciuti G (2015) Design and development of a soft robotic gripper for manipulation in minimally invasive surgery: a proof of concept. Meccanica 50:2855–2863

Author Index

K. L. Flores Rodríguez et al. (eds.), *Recent Trends in Sustainable Engineering*, Lecture Notes in Networks and Systems 297,
https://doi.org/10.1007/978-3-030-82064-0